Martina Leins

Waste Gas Decomposition by Plasmas

Martina Leins

Waste Gas Decomposition by Plasmas

Development and Spectroscopic Investigation of a Microwave-Generated Plasma Source at Atmospheric Pressure

Südwestdeutscher Verlag für Hochschulschriften

Impressum/Imprint (nur für Deutschland/ only for Germany)
Bibliografische Information der Deutschen Nationalbibliothek: Die Deutsche Nationalbibliothek verzeichnet diese Publikation in der Deutschen Nationalbibliografie; detaillierte bibliografische Daten sind im Internet über http://dnb.d-nb.de abrufbar.

Alle in diesem Buch genannten Marken und Produktnamen unterliegen warenzeichen-, marken- oder patentrechtlichem Schutz bzw. sind Warenzeichen oder eingetragene Warenzeichen der jeweiligen Inhaber. Die Wiedergabe von Marken, Produktnamen, Gebrauchsnamen, Handelsnamen, Warenbezeichnungen u.s.w. in diesem Werk berechtigt auch ohne besondere Kennzeichnung nicht zu der Annahme, dass solche Namen im Sinne der Warenzeichen- und Markenschutzgesetzgebung als frei zu betrachten wären und daher von jedermann benutzt werden dürften.

Verlag: Südwestdeutscher Verlag für Hochschulschriften Aktiengesellschaft & Co. KG
Dudweiler Landstr. 99, 66123 Saarbrücken, Deutschland
Telefon +49 681 37 20 271-1, Telefax +49 681 37 20 271-0
Email: info@svh-verlag.de
Zugl.: Stuttgart, Universität, Diss., 2010

Herstellung in Deutschland:
Schaltungsdienst Lange o.H.G., Berlin
Books on Demand GmbH, Norderstedt
Reha GmbH, Saarbrücken
Amazon Distribution GmbH, Leipzig
ISBN: 978-3-8381-1946-5

Imprint (only for USA, GB)
Bibliographic information published by the Deutsche Nationalbibliothek: The Deutsche Nationalbibliothek lists this publication in the Deutsche Nationalbibliografie; detailed bibliographic data are available in the Internet at http://dnb.d-nb.de.

Any brand names and product names mentioned in this book are subject to trademark, brand or patent protection and are trademarks or registered trademarks of their respective holders. The use of brand names, product names, common names, trade names, product descriptions etc. even without a particular marking in this works is in no way to be construed to mean that such names may be regarded as unrestricted in respect of trademark and brand protection legislation and could thus be used by anyone.

Publisher: Südwestdeutscher Verlag für Hochschulschriften Aktiengesellschaft & Co. KG
Dudweiler Landstr. 99, 66123 Saarbrücken, Germany
Phone +49 681 37 20 271-1, Fax +49 681 37 20 271-0
Email: info@svh-verlag.de

Printed in the U.S.A.
Printed in the U.K. by (see last page)
ISBN: 978-3-8381-1946-5

Copyright © 2010 by the author and Südwestdeutscher Verlag für Hochschulschriften Aktiengesellschaft & Co. KG and licensors
All rights reserved. Saarbrücken 2010

Contents

Contents		1
Introduction		7
1	**Fundamentals of Atmospheric Microwave Plasmas**	**11**
1.1	Plasma Parameters and General Plasma Properties	11
	1.1.1 Fundamental Plasma Parameters and Description of Plasmas	12
	1.1.2 Derived Plasma Parameters	16
1.2	Plasma and Electromagnetic Waves	18
1.3	Typical Plasma Parameters for Atmospheric Microwave-generated Plasmas	22
2	**Development of the Atmospheric Microwave Plasma Source (APS)**	**25**
2.1	Design of the Atmospheric Microwave Plasma Source APS	25
2.2	Analytical Solutions of the Maxwellian Equations for Cylindrical Resonators	29
2.3	Simulations with COMSOL Multiphysics^{TM}	31
	2.3.1 The Simulation Software COMSOL Multiphysics^{TM}	31
	2.3.1.1 The RF Module	32
	2.3.2 Simulation Results	34
	2.3.2.1 Simulations of Cylindrical Resonators Excited in Higher Modes	35
	2.3.2.2 Simulations of Configurations with Coupling Elements	37
	2.3.2.3 Simulation of the Realistic APS	41
2.4	Measurements of the Microwave Characteristics and Comparison to the Simulations	47
	2.4.1 Design and Construction of the E_{030}-Mode Resonator and the Coupling Elements	47
	2.4.2 Experimental Setup	49
	2.4.3 Experimental Results	51
3	**The Self Igniting APS**	**56**
3.1	Measurements of the Magnetron	56
3.2	Commonly used Experimental Setup for the Operation of Microwave-generated Plasma Sources	57
3.3	The Self Igniting E_{010}-APS	58
3.4	The Self Igniting E_{030}-APS	60

4 Characterisation of the Self Igniting APS — 65
4.1 The Experimental Setup — 65
4.1.1 Optical Emission Spectroscopy — 66
4.1.2 Decomposition of Waste Gases — 69
4.2 Characterisation of the APS Plasma — 70
4.2.1 Theoretical Descriptions of Atomic and Molecular Spectra — 71
4.2.1.1 Spectra of Atoms — 71
4.2.1.2 Spectra of Diatomic Molecules — 74
4.2.2 Overview Spectra and Applied Spectroscopic Methods — 79
4.2.3 Experimental Results and Interpretations — 84
4.3 Decomposition of Waste Gases — 107
4.3.1 Abatement of Volatile Organic Compounds — 107
4.3.2 Abatement of Perfluorinated Compounds — 113

5 Summary and Conclusions — 119

List of Figures — 124

List of Tables — 126

Bibliography — 127

Kurzfassung

Im Hinblick auf den Klimawandel nimmt die Reinhaltung der Luft einen immer höheren Stellenwert ein. Daher sind heutzutage der Abbau und die Reinigung von Abgasen eine der wichtigsten Herausforderungen der Menschheit geworden. Schädliche Gase hierfür sind unter anderem flüchtige organische Verbindungen (engl. volatile organic compounds (VOC)). Weitaus gefährlichere Gase stellen perfluorierte Verbindungen (engl. perfluorinated compounds (PFC)) dar, die ein 7000 − 23000-fach höheres klimaschädigendes Potenzial verglichen mit CO_2 haben und in wachsenden Industriezweigen, wie der Produktion von Halbleiterbauteilen, als Ätzgase in großem Umfang eingesetzt werden. Konventionell werden diese Abgase thermisch in Öl- oder Erdgasbrennern zersetzt, obwohl diese schwierig zu handhaben sind und dabei zusätzlich Kohlenstoffoxide produziert werden. Plasmaprozesse stellen hierfür eine viel versprechende Alternative dar, da keine zusätzlichen Kohlenstoffoxide entstehen. Hochfrequenz- und Gleichstromentladungen haben den Nachteil, dass sie Elektroden benötigen, die leicht erodieren, wenn sie in Berührung mit den Abgasen kommen. Daher bieten elektrodenlose Mikrowellenentladungen eine hervorragende Alternative für den Abbau von Abgasen.

Die vorliegende Arbeit befasst sich mit der Entwicklung und spektroskopischen Untersuchung einer bei $2,45\,\mathrm{GHz}$ mikrowellen-getriebenen Plasmaquelle bei Atmosphärendruck (engl. atmospheric pressure microwave plasma source (APS)) für die Abgasreinigung. Die Plasmaquelle beruht auf einem axial symmetrischen Hohlraum. Das Plasma wird durch ein Quarzrohr eingeschlossen, und die Gaszuführung erfolgt über eine metallische Düse. Für eine erfolgreiche Anwendung in industriellen Prozessen sind sowohl ein einfacher Zündvorgang als auch ein stabiler Betrieb unabdingbar. Um zu gewährleisten, dass das Plasma ohne weitere Zündhilfe gezündet werden kann, ist die detaillierte Kenntnis der elektrischen Feldverteilung erforderlich. Daher wurden finite Element-Simulation der elektrischen Feldverteilung mit der Software COMSOL Multiphysics^{TM} durchgeführt. Die Simulationsergebnisse konnten mit Hilfe eines Netzwerkanalysators verifiziert werden. Die Simulationen in Kombination mit den Messungen führten zu einer Konfiguration, die sowohl die Zündung ohne weitere Zündhilfe als auch einen stabilen Betrieb des Plasmas gewährleistet.

Die Charakterisierung des Plasmas für verschiedene Mikrowellenleistungen und Gasflüsse erfolgte mittels optischer Emissionsspektroskopie. Die Gastemperatur wurde mit Hilfe des $A^2\Sigma^+ - X^2\Pi_\gamma$-Übergangs im freien OH-Radikal ermittelt, während die Elektronentemperatur mittels eines Boltzmannplots aus zwei Sauerstoffatomlinien abgeschätzt werden konnte. Die Neutralteilchen- und Elektronendichte wurden aus diesen Temperaturen berechnet.

Des Weiteren wurde der Abbau von VOC am Beispiel von Propan und Toluol in Luftplasmen und von PFC am Beispiel von CF_4 und SF_6 in Stickstoffplasmen untersucht. Die Analyse der Roh- und Reingase erfolgte mittels Fourier-Transform-Infrarot-Spektroskopie, einem Flammenionisationsdetektor, einem Quadrupolmassenspektrometer und einem Gaschromatographen. Die Messungen ergaben, dass obwohl eine hohe Abbaueffizienz von über 99 % für die Zersetzung von VOC erzielt werden konnte, der Einsatz der APS für die Reinigung von VOC fragwürdig ist, da kritische Nebenprodukte und große Mengen an Stickoxiden erzeugt wurden. Andererseits zeigten die Messungen auch, dass PFC vollständig und ohne dass kritische Nebenprodukte entstehen, abgebaut werden können und sich daher die APS sehr gut für den Abbau von PFC eignet.

Abstract

In view of the world climate change the cleaning and purification of waste gases has become one of the most urgent tasks for humankind nowadays. Harmful gases are volatile organic compounds (VOC). However, even more hazardous gases are perfluorinated compounds (PFC) which have a 7000..23000 times higher green house potential compared to CO_2 and are widely used for etching processes in growing industry sectors like the production of semiconductors. Conventionally these waste gases are treated in oil or gas combustions which are complex to handle and produce additional carbon oxides. Hot plasma processes offer a promising alternative for this purpose, since the production of additional carbon oxides is prevented. Common RF and DC discharges have the disadvantage of electrodes which would erode when they come in contact with the waste gases. An excellent option is provided by an electrodeless microwave plasma torch at atmospheric pressure. This work deals with the development and spectroscopic study of a plasma source at atmospheric pressure powered by 2.45 GHz microwaves (APS) for the abatement of waste gases. The plasma source is based on an axially symmetric cavity. The plasma is confined in a quartz tube and a metallic nozzle is used for the gas inlet. For a successful application in industry simple ignition of the plasma as well as stable plasma operation are indispensable. To guarantee that the plasma can be ignited without any additional igniters detailed information about the electric field distribution is required. Therefore, finite element simulations of the electric field were conducted by using the simulation software COMSOL Multiphysics^{TM}. The simulation results were verified by measurements with a network analyser. The simulations combined with the measurements led to a configuration which provides an ignition of the plasma without any additional igniters as well as stable plasma operation.

The characterisation of the plasma was performed by means of optical emission spectroscopy for different microwave powers and air flows. The gas temperature was measured by using the $A^2\Sigma^+ - X^2\Pi_\gamma$-transition of the free OH radical while the electron temperature was estimated from a Boltzmann plot of two atomic oxygen lines. The neutral particle and electron density were calculated from these temperatures.

Furthermore, the decomposition of as exemplary VOC propane and toluene in air plasmas and as exemplary PFC CF_4 and SF_6 in nitrogen plasmas was studied. The analyses of the raw and clean gases were performed using Fourier-Transform Infra-Red spectroscopy, a flame ionisation detector, a quadrupole mass spectrometer, and a gas phase chromatograph. The measurements revealed that the suitability for the abatement of VOC is questionable even though destruction and removal efficiencies of over 99 % are reached since critical by-products and large amounts of NO_x are produced. However, the measurements also showed that the PFC can be completely decomposed and that no critical by-products are formed and therefore the APS is well suited for the abatement of PFC.

Introduction

Climate scientists as well as scientific associations agree that one of the most important causes of the global warming since the beginning of the industrialisation is the aerating of greenhouse gases which are produced by humankind. Effects of the global warming can already be observed today. The increase of the global average temperature in the ground level of 0.74 °C between 1906 and 2005 is the main evidence of the world climate change. The rising of the sea level, the melting of the glaciers and weather changes such as drought periods or heat-waves can be adduced as further effects of the global warming. These effects have remarkable implications on the human security, health, economy and environment. So stopping the world climate change or at least attenuating its effects is one of the biggest tasks of humankind nowadays. As a result in 1997 the United Nations decided in Kyoto an additional protocol to the United Nations Framework Convention on Climate Change (UNFCCC), the so called Kyoto Protocol [1]. This Kyoto Protocol became operative in February 2005 and ends 2012 and prescribes binding upper values for the emission of greenhouse gases. Regulated gases are carbon dioxide (CO_2), methane (CH_4), dinitrogenoxide (N_2O), volatile organic compounds (VOC), and perfluorinated compounds (PFC). PFC are widely used for etching processes in growing industry sectors such as semiconductor industries and in thin-film technologies. Often used PCF like CF_4 or SF_6 have 7000..23000 times higher greenhouse potentials compared to CO_2. So the reduction of exhaust gases and the cleaning and purification of waste gases is becoming a more and more important task for enterprises. Conventionally these VOC and PFC are treated thermally in oil or gas combustions even though more CO_2 is produced. Since oil or gas combustions already produce waste gases by themselfs, for example carbon and nitride oxides, and since they are complex to handle and often need a long time to get started, they are only reasonable for huge and steady waste gas flows.

Hot (thermal) plasma processes at atmospheric pressure, like plasma torches, offer a promising alternative for this purpose, since these plasma sources provide high electron, ion, and radical densities and no or only very few additional waste gases are produced. Common RF- or DC-plasma torches have the disadvantage of electrodes which can erode when they get in contact with the plasma or the waste gases itself. An electrodeless microwave-generated plasma torch at atmospheric pressure provides an excellent option for this purpose.

Beside the purification of waste gases further possible applications of the microwave plasma torch are other chemical syntheses, for example the pyrolysis of methane to carbon and hydrogen which has an application in crewed space flights. Another application is the treatment of surfaces, for example the activation to increase the adhesion of lacquers or glue.

Contemporary used microwave plasma sources at a frequency of 2.45 GHz are for example the surfatron [2, 3, 4], waveguide-based axial-type microwave plasma sources [5, 6, 7, 8, 9], (tapered) waveguide-based systems [10, 11, 12, 13, 14, 15] and resonator based microwave plasma sources [16, 17, 18, 19]. Some of them have recently been used to study the decomposition of waste gases [16, 17, 18, 19, 20, 21, 22, 23, 24] but these sources have disadvantages. The surfaguide [25] for example needs complex cooling systems of the discharge tube and the plasma undergoes filamentation so not the whole gas flow is treated [24, 26, 27]. An even profounder problem of these sources is that for the plasma ignition additional igniters such as AC electric sparks or arc torch modules [20, 21, 22, 28, 29] are needed. On the other hand, when no igniters are necessary, only pulsed plasma operation is possible [16, 17, 18, 19]. The utilized sparks or arc modules are susceptible to erode when they get in contact with the waste gases and/or the plasma.

Due to one of the most important tasks nowadays of cleaning and purification of waste gases the development of an atmospheric microwave plasma source for this purpose is indispensable. However, for a successful application in industrial processes it is necessary that the ignition of the plasma is as simple as possible and always guaranteed. Furthermore, for a straight forward handling of the plasma source and a complete decomposition of the waste gases, a stable and continuous plasma operation as well as an efficient absorption of the supplied microwave power is indispensable. In addition the ability to treat large gas flows is required. However, all of the contemporary atmospheric microwave plasma sources have at least one of the disadvantages mentioned above and therefore are inappropriate for the successful abatement of waste gases.

Thus in view of the important task of purifying waste gases the development of an atmospheric microwave plasma source with the properties above mentioned is indispensable and the first task of the present work.

Furthermore, to obtain a better understanding of the purification of waste gases and to influence the decomposition of these gases and the formation of by-products, information about the possible reaction channels is needed. Therefore, knowledge about the species in the plasma and their temperatures as well as their densities is necessary. Pure noble gas or noble gas dominated microwave-sustained discharges at atmospheric pressure have already been widely explored [7, 8, 9, 12, 13, 30, 31, 32, 33, 34, 35, 36, 37]. However, noble gas or noble gas dominated discharges differ from discharges in molecular gases like air, O_2-, N_2-, or air or N_2-plasmas containing VOC or PFC. Molecular gas microwave plasma discharges at atmospheric pressure have not been thoroughly and incompletely characterised [6, 10, 14, 19]. Therefore, the characterisation of waste purification relevant plasmas and a detailed characterisation of the plasmas produced by this developed microwave plasma torch is essential.

At last the suitability for the abatement of waste gases of this atmospheric microwave plasma torch has to be analysed.

Since an excellent option for the decomposition of waste gases is provided by electrodeless atmospheric microwave plasma sources the fundamentals of these plasmas are presented in chapter 1.

Thereafter, the development of the atmospheric microwave plasma source (APS) is presented in chapter 2. For a successful operation in the industry an easy ignition as well as stable plasma operation is indispensable and therefore the APS has to provide all these properties. To guarantee that an electric field strength which is high enough to ignite plasma without any additional igniters is reached, detailed information about the electric field distribution is needed. The Maxwellian equations can be solved analytically for simple configurations which is shown in section 2.2. However, information about the electric field distribution for more complex configurations can only be obtained via finite element simulations. Therefore, simulations of the electric field as well as calculations of the Eigenfrequency with the software COMSOL Multiphysics^{TM} were performed and are described in section 2.3. The simulations led to configurations which provide high electric fields. These configurations were realised, then measured with a network analyser, and compared to the simulations which are presented in section 2.4.

The simulations in combination with the measurements led to a configuration which is able to ignite plasma without any additional igniters and maintains stable plasma operation. A detailed description of this configuration is given in chapter 3.

After that the characterisation of the APS plasmas as well as studies concerning the abatement of waste gases are presented in chapter 4.

Since gas temperatures of about 500 K..10000 K are expected and reactive gases will be used, Langmuir probes, which are often used to determine electron temperatures and densities, would erode and maybe produce misleading results. These metallic Langmuir probes would also affect the electric field distribution and thus disturbe the discharge. Optical emission spectroscopy (OES) represents a non-disturbing diagnostic. Furthermore, using OES nearly all species which are present in the plasma can be identified and gas rotational T_{rot}, vibrational T_{vib}, excitation T_{ex}, and electron T_e temperatures as well as densities can be measured. Since this work focuses on the characterisation of waste-purification-relevant plasmas, air plasmas were characterised. To measure the gas rotational temperature T_{rot}, the $A^2\Sigma^+$ - $X^2\Pi_\gamma$-transition of the free OH radical was used and therefore humid air plasmas were analysed. The rotational temperature provides a good estimation of the translation or gas temperature. Atomic lines can be used to determine the excitation temperature T_{ex} which displays the electron temperature T_e when a Maxwellian velocity distribution and a Boltzmann population are existent. In air plasmas two atomic oxygen lines could be observed which were used to determine T_{ex}. The validity of the Boltzmann plot of only two atomic oxygen lines was verified by Boltzmann plots of more oxygen lines, which were observed in oxygen plasmas. The electron density n_e for air plasmas could be rated assuming partial local thermodynamic equilibrium and by using Saha's equation. These results are described in section 4.2.

The last part of this work deals with the degradation of pollutants in waste gases such as air flows containing VOC or nitrogen flows containing PFC. Therefore, air flows containing propane and toluene and nitrogen flows containing CF_4 and SF_6 were treated with the APS and the raw and clean gases were analysed by using Fourier-Transform Infra-Red spectroscopy (FT-IR), Quadrupolemass spectroscopy and a flame ionisation detector (FID) as well as a gas-phase chromatograph to measure the destruction and removal efficiency (DRE) and to obtain information about reaction and by-products. The plasma itself was again characterised by optical emission spectroscopy which gave information about the species present in these plasmas. Results of the analyses of the raw and clean gases and of the characterisation of the plasma allowed to draw conclusions about possible reaction channels which are presented in section 4.3.

The last chapter 5 provides the summary and conclusions.

Chapter 1

Fundamentals of Atmospheric Microwave Plasmas

Conventionally waste gases are degraded thermally by oil or gas combustion which intrinsicly produces carbon oxides. Furthermore, these combustion processes are complex, difficult to handle, need a long time to get started, and therefore only a permanent operation is reasonable. Hot (thermal) plasma processes at atmospheric pressure provide a promising alternative to the conventional treatment of waste gases, since they offer high electron, ion and radical densities. Furthermore, the intrinsic formation of carbon oxides is prevented. Common DC- or RF-discharges have the disadvantage of electrodes which erode when they come in contact with the reactive waste gases. A microwave plasma, which has no electrodes, offers an excellent option for this purpose. For a comprehensive understanding of the processes involved in the decomposition of waste gases knowledge of the fundamentals of atmospheric pressure microwave plasmas is essential. This chapter provides an introduction to these fundamentals.

1.1 Plasma Parameters and General Plasma Properties

Starting from the three well known aggregate states solid, liquid, and gaseous, increasing the temperature plasmas follow after the gaseous state and are sometimes named as the fourth aggregate state. Thus plasma physics covers the generation, characterisation, and description of the properties of ionized gases. In contrast to normal gases plasmas also consist - additionally to neutrals - of ions and electrons. The charged particles lead to the circumstance that plasmas behave as a well conducting fluid, which interacts with electromagnetic fields. Since plasmas have these long-range electromagnetic interactions, plasmas show collective behaviour like for example oscillations and waves.

A first classification of plasmas can be performed by the temperature and density of the electrons. The electron density n_e ranges over many decades and lasts from very thin gases in the interstellar space with an electron density of about $n_e \approx 10^5 \, \text{m}^{-3}$ to plasmas in stars with electron densities of about $n_e \approx 10^{30} \, \text{m}^{-3}$. The electron temperature T_e ranges from room temperature for some technical plasmas to several million Kelvin in stars. Fig. 1.1 gives an overview of the variety

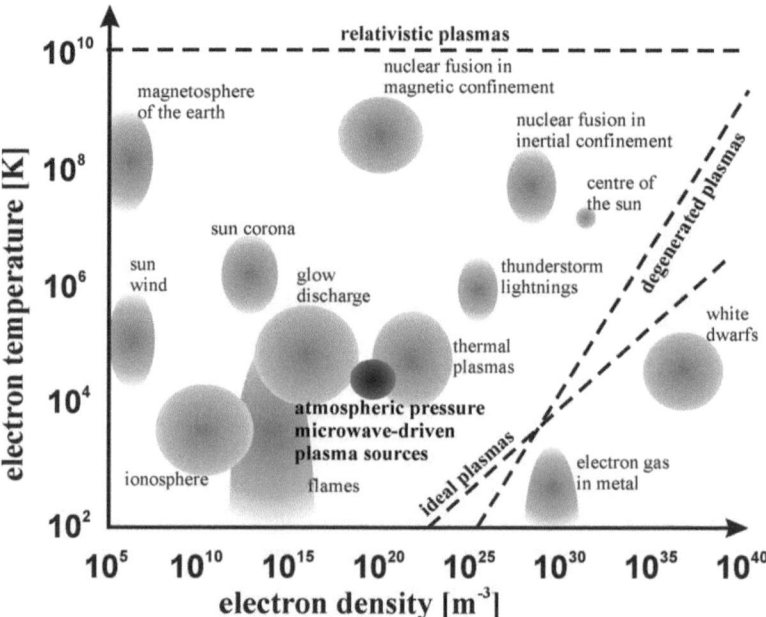

Figure 1.1: Overview of the variety of different plasmas. The atmospheric pressure microwave-generated plasmas can be found between the glow discharges and thermal plasmas.

of different plasmas and classifies the atmospheric pressure microwave-generated plasmas in this diagram. These plasmas are located between the glow discharges and thermal plasmas.

1.1.1 Fundamental Plasma Parameters and Description of Plasmas

A plasma consists of many different particles, molecules, radicals, atoms, ions and electrons. The entire density of the plasma can be split into the density of the single species. Usually, one distinguishes between the density of the neutrals n_a, the density of the ions n_i, and the electron density n_e. These various densities partly characterise the plasma and therefore the measurement of these densities is often of great interest.

However, the plasma is not only characterised by the density of each component but also by the velocity of the particles. If the plasma is in complete thermodynamic equilibrium, which will be described below, the velocity distribution of the particles is chaotic and a temperature can be defined which characterises the plasma. Laboratory plasmas are not in complete thermodynamic equilibrium but often they can be approximated by other models, which are described in the following. Most of these models lead to more than only one temperature.

The densities as well as the temperatures are then designated as plasma parameters.

Complete Thermodynamic Equilibrium

If the plasma is in complete thermodynamic equilibrium it can be described by a few macroscopic quantities like the temperature and pressure. Then the densities of neutrals, ions, and electrons are only dependent on the temperature T and pressure p. All elementary processes such as ionisation or excitation, are in equilibrium with their corresponding reversal processes, like the recombination or deexcitation and the complete energy transfer between the particles is ensured. The plasma is uniquely described by the Saha-Eggert-equation [45]:

$$\frac{n_e n_i}{n_a} = \frac{2(2\pi m_e k_b T)^{3/2}}{h^3} \frac{Q_i(T)}{Q_a(T)} e^{-\frac{E_i}{k_b T}}, \qquad (1.1)$$

Dalton's law:

$$p = (n_a + n_i + n_e) k_b T \qquad (1.2)$$

and the quasi neutrality, which will be explained in section 1.1.2:

$$n_i = n_e. \qquad (1.3)$$

Here only a single ionisation is assumed. Otherwise the corresponding sums must be calculated. In equation 1.1 Q_i and Q_a are the partition functions of the neutrals and ions and E_i is the ionisation energy. Furthermore, m_e is the electron mass, h Planck's quantum, and k_b the Boltzmann constant. The particles obey a Maxwellian velocity distribution what implies that the particle velocity has no preferred direction and the swirling of the particles is perfectly chaotic. The Maxwellian velocity distribution f_M is given by the following equation:

$$f_M(v, T, m) = \frac{4}{\sqrt{\pi}} \frac{v^2}{v_{th}^3} e^{-\frac{v^2}{v_{th}^2}}, \qquad (1.4)$$

where v_{th}^2 is the most probabilistic velocity:

$$v_{th} = \sqrt{\frac{2k_b T}{m}}. \qquad (1.5)$$

In complete thermodynamic equilibrium the population of the different atomic and ionic levels follows a Boltzmann distribution. Then the population density n_m of a certain energy level is given by:

$$n_m = n \frac{g_m}{Q} e^{-\frac{E_m}{k_b T}}. \qquad (1.6)$$

with g_m being the statistical weight. The radiation of a plasma, which is in complete thermodynamic equilibrium, is constituted by Kirchhoff-Planck's law, which describes the energy distribution of the electromagnetic radiation in dependence of the frequency ν [46]:

$$L_\nu = \frac{2h}{c^2} \nu^3 \left[e^{-\frac{h\nu}{k_b T}} - 1 \right]^{-1}. \qquad (1.7)$$

However, most laboratory plasmas are not in complete thermodynamic equilibrium but in other equilibria. Therefore, the local thermodynamic equilibrium, which is a first step to the description of laboratory plasmas, is described in the following.

Local Thermodynamic Equilibrium

If no complete thermodynamic equilibrium is present, often, when collisional processes predominate, the energy transfer processes make sure that locally the electrons as well as the other particles have a Maxwellian velocity distribution in good approximation with the same temperature. Then the ionisation and the excitation situation of the atoms and ions are still given by the Saha-Eggert equation and the Boltzmann distribution, respectively, at this temperature and the plasma is in local thermodynamic equilibrium [46]. To provide, that the energy transfer processes lead to a uniform temperature of the electrons and the other particles, the electron density must attain a critical value which is given by the following relation [45]:

$$n_e \gg 1.6 \cdot 10^{18} \sqrt{T} (\Delta E)^3 \, \text{m}^{-3}. \tag{1.8}$$

ΔE is the energy distance of the levels in which transitions of the regarded level can occur. Furthermore, a local thermodynamic equilibrium is only present when the mean free path length λ is distinctly smaller than the temperature and density gradient. This can be expressed in the two following relations [46]:

$$\lambda |\frac{\Delta T}{\Delta x}| \ll T \quad \text{and} \quad \lambda |\frac{\Delta n}{\Delta x}| \ll n. \tag{1.9}$$

In contrast to plasmas in complete thermodynamic equilibrium, the radiation from a plasma in local thermodynamic equilibrium is no longer given by the Kirchhoff-Planck's law. The radiation is now, besides the temperature, defined by the population densities as well as by the complex quantum mechanical structure of the particles. Therefore, the radiation from a plasma which is in local thermodynamic equilibrium provides much more information about the plasma and its constituents. Plasmas in local thermodynamic equilibrium are aspired but are only found in exceptional cases like in arc plasmas with $n_e \approx 10^{22}..10^{23} \, \text{m}^{-3}$. Then the local thermodynamic equilibrium provides a good description for these plasmas [48]. However, often the partial local thermodynamic equilibrium provides a better description of laboratory plasmas than the local thermodynamic equilibrium and therefore the partial local thermodynamic equilibrium is described in the following part.

Partial Local Thermodynamic Equilibrium

Often the requirement in equation 1.8 is not fulfilled. Commonly the plasma is heated by a high frequently electromagnetic field and only the light electrons can gather energy from this field. The heavy particles are heated by collisions with the fast electrons but the energy transfer is small due to the mass ratio of the heavy particles to the electrons and two velocities distributions for the electrons and heavy particles develop. Then, instead of a uniform temperature for all particles, the particles can have different temperatures for their Maxwellian velocity distributions. It must be noted, that generally the electron temperature T_e is higher than the temperature of the heavy particles. Furthermore, an ionisation temperature for each ionisation state as well as for every population density an excitation temperature can be associated. A partial local thermodynamic equilibrium is present if the excitation temperature of all energy levels, except the one of the ground state, coincides with the electron temperature [46]. Since the population of the electronic excited states is induced by electron collisions in most laboratory plasmas, but the deexcitation on the other hand, beside the electron collisionional deexcitations, is also caused by spontaneous emission, which is not balanced, the excited states are underpopulated compared to the ground state. Therefore, equation 1.6 must be corrected by a factor $b_m \leq 1$ [48]. Furthermore, the Saha-Eggert equation 1.1 must also be corrected by a factor $\frac{1}{a}$, with $a \geq 1$, since the ionisation is smaller than in plasmas which are in complete thermodynamic equilibrium [48]. Moreover, Dalton's law 1.2 must be extended to:

$$p = n_e k_b T_e + (n_a + n_i) k_b T_g, \tag{1.10}$$

where T_e is the electron temperature and T_g is the temperature of the heavy particles, the gas temperature [48]. For plasmas in partial local thermodynamic equilibrium the ionisation and the excitation is controlled by the electron temperature T_e, the dissociation of molecules by the temperature of the heavy particles T_g, the rotational excitations of molecules by T_g, and the vibrational excitation of molecules by T_e and T_g.

Corona Equilibrium

For plasmas with low electron densities ($n_e < 10^{17}\,\text{m}^{-3}..10^{19}\,\text{m}^{-3}$) the deexcitational electron collisions can be neglected against the spontaneous emission. The model assumes that the excitation is only controlled by electron collision whereas the deexcitation is given by spontaneous emission and photo recombination. This is, for example found in the plasma of the sun's corona and therefore the model which describes such plasmas is called corona model.

Collisional-Radiative Models

For plasmas which are located between the two regimes with electron densities $10^{19}\,\text{m}^{-3} \geqslant n_e \geqslant 10^{22}\,\text{m}^{-3}$, which are described by the partial local thermodynamic equilibrium and the corona model, the collisional-radiative model provides a good description. A rate equation is formulated in this model, which respects all excitation and deexcitation processes between the excited state and the ground state. Typically, electron collision excitation and deexcitation as well as photo excitation and absorption and spontaneous as well as induced emission are considered. Collisional-radiative models can also be used to calculate the constants a and b, which were introduced for the description of the partial local equilibrium.

Different kinds of plasmas can be described with the models presented above even though not all plasmas can be described by one of these models. Typical plasma parameters of atmospheric pressure microwave-generated plasmas will be presented in section 1.3 and then the model which suits best will be figured out and it will be discussed why a certain model was chosen to describe the plasma of the APS.

1.1.2 Derived Plasma Parameters

The variety of all the plasmas presented in Fig. 1.1 can be described by the same parameters. These parameters can be derived from the plasma parameters described in the previous section 1.1.1 and will be presented in the following.

Quasi Neutrality

As already mentioned a plasma consists of neutral and charged particles. However, the plasma is quasi neutral on macroscopic scales. This implies that the numbers of singly ionized positively charged particles and of negatively charged particles must be equal. For singly charged ions this can be expressed by the following equation:

$$n_e = n_i = \frac{n}{2}, \tag{1.11}$$

where n_i is the ion density and n is the plasma density.

Degree of Ionisation χ

Since a plasma, among neutral particles, also consists of charged particles, a degree of ionisation χ can be described by the ratio of the density of the charged particles n_i to the density of the uncharged particles n_a:

$$\chi = \frac{n_e}{n_a + n_e} \approx \frac{n_e}{n_a} = \frac{n_i}{n_a}. \tag{1.12}$$

The degree of ionisation ranges from almost complete ionised fusion plasmas, which have an ionisation degree of nearly $\chi = 1$, to technical plasmas, which often have distinct lower ionisation degrees of about $\chi \approx 10^{-6}..10^{-2}$ and $\chi \approx 10^{-8}$ in extreme cases.

Debye Length λ_D

The quasi neutrality is not retained on the microscopic scale, since ions reject other ions whereas electrons are attracted. Therefore, the ions are enclosed by electrons to shield their own charge. On the other hand, the electrons are enclosed by ions for the same purpose. Hence the electric

1.1. PLASMA PARAMETERS AND GENERAL PLASMA PROPERTIES

potential is not given by the electric potential in vacuum but by the Debye-Hückel potential:

$$\phi(r) = \frac{q}{4\pi\epsilon_0} \cdot \frac{1}{r} e^{-\frac{\sqrt{2}r}{\lambda_D}}, \qquad (1.13)$$

where q is the charge of the particle, r the distance from the charged particle, and λ_D the so called Debye length. Thus the electric potential of a charged particle in a plasma decreases faster than in vacuum. The Debye length is given by the following equation

$$\lambda_D = \sqrt{\frac{\epsilon_0 k_b T}{n e^2}}, \qquad (1.14)$$

with k_b being the Boltzmann constant and T the temperature [42]. In a plasma a charged particle is enclosed by a cloud with oppositely charged particles whose dimension is of about the Debye length λ_D.

With the Debye length the quasi neutrality of the plasma on the macroscopic scale can then be expressed by the following relation:

$$L \gg \lambda_D \qquad (1.15)$$

with L being the dimension of the plasma. This illustrates that plasmas are quasi neutral when they are regarded from the outside.

Plasma Parameter N_D

It must furthermore be assured that enough particles are located in this charge cloud. The number of particles in the Debye sphere is given as follows:

$$N_D = n \cdot \frac{4}{3}\pi \lambda_D^3 = \frac{4\pi}{3} \cdot \left(\frac{\epsilon_0 k_b T}{e^2}\right)^{3/2} \cdot n^{-1/2} \qquad (1.16)$$

and is defined as the plasma parameter N_D [42].

Plasma Frequency ω_p

The collective behaviour is expressed by plasma oscillations and waves. The simplest plasma oscillation is an oscillation where the electrons oscillate versus the ions. The frequency of this oscillation is given by the electron plasma frequency:

$$\omega_{pe} = \sqrt{\frac{ne^2}{\epsilon_0 m_e}}. \qquad (1.17)$$

The electron plasma frequency ω_{pe} is of major importance to the propagation of waves in plasmas.

The whole plasma frequency ω_p is given by:

$$\omega_p^2 = \omega_{pe}^2 + \omega_{pi}^2 \qquad (1.18)$$

where $\omega_{pi} = \sqrt{\frac{nZ^2e^2}{\epsilon_0 m_i}}$ is the ion plasma frequency. However, the plasma frequency is dominated by the electron plasma frequency, due to the mass difference of ions and electrons and therefore the plasma frequency can be approximated by the electron plasma frequency: $\omega_p \approx \omega_{pe}$.

1.2 Plasma and Electromagnetic Waves

Since the analysed plasma source is generated by microwaves and therefore the energy to the plasma is supplied via a high frequency electromagnetic field, the behaviour of waves in unmagnetized, temperated plasmas will be discussed in the following. The plasma will be described by the Drude-Model and the conductivity, the penetration depth, and the absorbed power in a collision dominated plasma will be deduced [44].

In this model the current is entirely provided by the electrons since only the electrons are able to follow the high frequency field at the excitation frequency of 2.45 GHz. Furthermore, the model is limited to weak ionised plasma, without a pressure gradient Δp, and without any external magnetic field. All collisions are assumed to be elastic, so that this model does not incorporate excitation and ionisation. Since only weak ionised plasmas are described the electron neutral particle collision frequency ν_{en} is applied for the collision frequency.

Conductivity σ

At first the plasma can be described by Ohm's law, which expresses the behaviour of the charged particles, which can follow the electric field, here the electrons [44]:

$$\nu_{en} j + \frac{dj}{dt} = \frac{n_e e^2}{m_e} E = \epsilon_0 \omega_{pe}^2 E, \qquad (1.19)$$

with j being the electron current density and ω the angular frequency of the irradiated high frequency wave. On the left side the loss term of the electron current density, which is determined by the collisions with the frequency ν_{en}, as well as the temporal derivative of j are given while the right side describes the term, which is given by the electric field E. If a harmonic time dependence $e^{i\omega t}$ of the electric field and the electron current density is assumed, the time derivative of the electron current density can be replaced by $i\omega$ and equation 1.19 becomes:

$$\nu_{en} j + i\omega j = \frac{n_e e^2}{m_e} E. \qquad (1.20)$$

1.2. PLASMA AND ELECTROMAGNETIC WAVES

The electric field is generally linked to the conductivity σ by following relation [44]:

$$j = \sigma E. \tag{1.21}$$

When equation 1.20 is transposed

$$j = \frac{n_e e^2}{m_e} \frac{1}{\nu_{en} + i\omega} E \tag{1.22}$$

and compared to 1.21 the conductivity is given by:

$$\sigma = \frac{n_e e^2}{m_e} \frac{1}{\nu_{en} + i\omega}. \tag{1.23}$$

The behaviour of this equation becomes more apparent when the normalised collision frequency $\frac{\nu_{en}}{\omega}$ is used and when the equation is separated into a real and an imaginary part [44]:

$$\sigma = \frac{n_e e^2}{m_e \nu_{en}} \frac{\left(\frac{\nu_{en}}{\omega}\right)^2 - i\frac{\nu_{en}}{\omega}}{1 + \left(\frac{\nu_{en}}{\omega}\right)^2} \tag{1.24}$$

To obtain the total conductivity equation 1.24 must be complemented by the vacuum displacement current and becomes [44]:

$$\sigma_{tot} = \frac{n_e e^2}{m_e \nu_{en}} \frac{\left(\frac{\nu_{en}}{\omega}\right)^2 - i\frac{\nu_{en}}{\omega}}{1 + \left(\frac{\nu_{en}}{\omega}\right)^2} + j\omega\epsilon_0. \tag{1.25}$$

The real part displays the Ohmic losses while the imaginary part leads to an inductive behaviour of the plasma. The real part grows quadratically in dependence of the collision frequency and therefore the real part dominates in collision dominated plasma.

Relative Permittivity ϵ_r and Refraction Index n_{ref}

The relative permittivity ϵ_r of the plasma can be deduced from the total conductivity given in equation 1.25

$$\epsilon_r = 1 - i\frac{\sigma}{\epsilon_0 \omega}, \tag{1.26}$$

and with the plasma frequency ω_p one gets:

$$\epsilon_r = 1 - \frac{\omega_p^2}{\omega(\omega - i\nu_{en})}. \tag{1.27}$$

The refraction index n_{ref} is given by the root of the relative permittivity ϵ_r:

$$n_{ref} = \sqrt{\epsilon} = \mathrm{Re}(n_{ref}) + i\mathrm{Im}(n_{ref}) = n_{Re} + in_{Im} \tag{1.28}$$

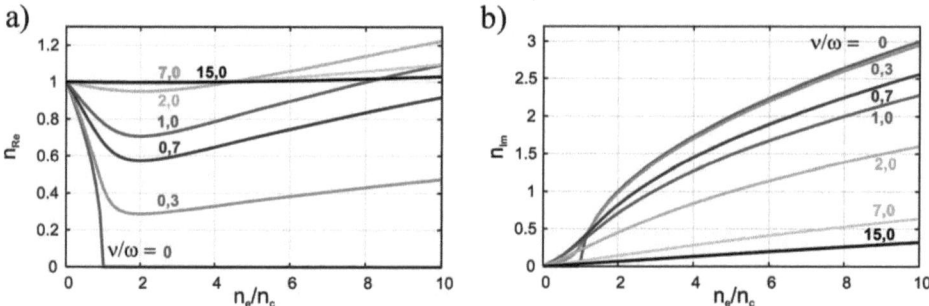

Figure 1.2: Dependence of the real and imaginary part of the refraction index on $\frac{n_e}{n_c}$ for different collision frequencies $\frac{\nu_{en}}{\omega}$. If $\frac{\nu_{en}}{\omega}$ is zero the real part of the refraction index becomes zero for $\frac{n_e}{n_c} = 1$. A wave cannot propagate. However, for high densities and large collision frequencies the refraction index can become bigger than $n_{ref} = 1$ and a wave can penetrate into the plasma [41].

with [44]

$$n_{Re} = \sqrt{\frac{1}{2}(1-\alpha) + \frac{1}{2}\sqrt{(1-\alpha)^2 + \left(\frac{\nu_{en}}{\omega}\alpha\right)^2}}, \quad (1.29)$$

$$n_{Im} = \sqrt{-\frac{1}{2}(1-\alpha) + \frac{1}{2}\sqrt{(1-\alpha)^2 + \left(\frac{\nu_{en}}{\omega}\alpha\right)^2}}, \quad (1.30)$$

$$\alpha = \frac{\left(\frac{\omega_p}{\omega}\right)^2}{1 + \left(\frac{\nu_{en}}{\omega}\right)^2}. \quad (1.31)$$

If the exciting frequency ω gets smaller than the plasma frequency ω_p and if no collisions ($\nu_{ne} = 0$) are considered, the refraction index n_{ref} becomes zero: $n_{ref} \to 0$. Then no waves can propagate in the plasma. The electron density at which the exciting wave can no longer penetrate into the plasma is reached when ω is equal to the plasma frequency ω_p and a cutoff density n_c can be defined:

$$n_c = \frac{\epsilon_0 m_e}{e^2}\omega^2. \quad (1.32)$$

The situation changes if collisions are considered. Even for $\omega < \omega_p$ or $n_e > n_c$, the real part of the refraction index n_{ref} can be greater than one and a wave with a frequency of $\omega < \omega_p$ can penetrate into the plasma. Fig. 1.2 shows in a) the real part of the refraction index in dependence of $\frac{n_e}{n_c}$ while b) shows the same for the imaginary part for different collision frequencies $\frac{\nu_{en}}{\omega}$.

Skin Depth δ of the Wave

How deep the wave penetrates into the plasma is described by the skin depth δ. The skin depth is defined as the length when the amplitude of the penetrated wave has dropped to $1/e$. This can

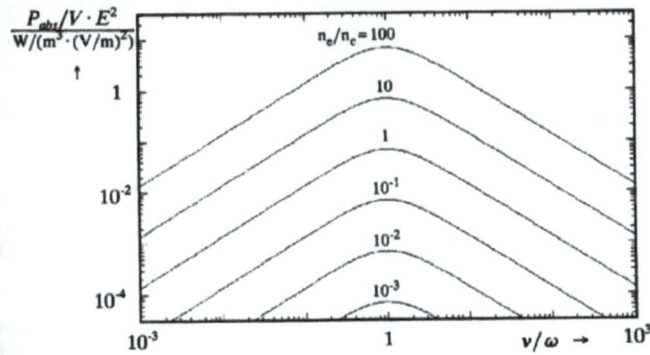

Figure 1.3: The relative power absorption as function of the collision frequency ν_{en} normalised to ω for different $\frac{n_e}{n_c}$ ratios [44].

be calculated from the relation $k = \frac{\omega}{c} n_{ref}$. Hereby the imaginary part causes the losses:

$$\delta = \frac{c}{\omega} \frac{1}{n_{Im}}. \tag{1.33}$$

So for smaller frequencies ω the skin depth is relatively large for the same refraction index n_{ref}.

Power Absorption by the Plasma

The power P, which is absorbed by the plasma per volume V can also be deduced from this model and is given by:

$$p = \frac{P}{V} = j \cdot E = \sigma E^2. \tag{1.34}$$

Therefore, the effective power per volume p_w is the real part of the entire power density [44]:

$$p_w = \frac{n_e e^2}{m_e \nu_{en}} \frac{\left(\frac{\nu_{en}}{\omega}\right)^2}{1 + \left(\frac{\nu_{en}}{\omega}\right)^2} E^2 \tag{1.35}$$

$$\frac{P_w}{VE^2} = \frac{n_e}{n_c} \epsilon_0 \omega \frac{\frac{\nu_{en}}{\omega}}{1 + \left(\frac{\nu_{en}}{\omega}\right)^2} \tag{1.36}$$

Fig. 1.3 shows the power absorption in dependence of the collision frequency $\frac{\nu_{en}}{\omega}$ for different $\frac{n_e}{n_c}$ ratios and for a microwave frequency of 2.45 GHz. It can be seen that the most power is absorbed when the frequency of the microwave ω is of about the same order of magnitude as the collision frequency ν_{en}.

1.3 Typical Plasma Parameters for Atmospheric Microwave-generated Plasmas

The fundamental and derived plasma parameters of atmospheric pressure microwave-generated plasmas are summarised in table 1.1. A plasma is classified as an ideal plasma if [42]:

$$\lambda_D \ll L,$$
$$N_D \gg 1 \quad \text{and} \tag{1.37}$$
$$\omega_p \cdot \frac{1}{\nu_{en}} > 1.$$

Since the Debye length of an atmospheric pressure microwave-generated plasma is about $\lambda_D \approx 0.1..1\,\mu\text{m}$ and the dimension of the APS plasma L is within centimetres, the first requirement in 1.37 is fulfilled. With N_D greater than 40 the second requirement, that the number of particles in the Debye sphere should be much larger compared to one, is not that well provided as in low pressure or fusion plasmas with N_D of about $2 \cdot 10^4$ and $2 \cdot 10^8$, respectively, [43]. $\omega_p \cdot \frac{1}{\nu_{en}} = 8..90$ and thus is greater than one. So the criteria for an ideal plasma are tenably fulfilled for an atmospheric pressure microwave-generated plasma.

The electron density n_e of an atmospheric pressure microwave-generated plasma ranges between $n_e \approx 10^2..10^{21}\,\text{m}^{-3}$ while the cutoff density for a microwave with a frequency of $2.45\,\text{GHz}$, $\omega = 2\pi \cdot 2.45\,\text{GHz} = 1.54 \cdot 10^{10}\,\text{s}^{-1}$, is already reached at a value of $n_c = 7.4 \cdot 10^{16}\,\text{m}^{-3}$, as can be seen in table 1.1. Thus the quotient of $\frac{n_e}{n_c} = \frac{\omega_p^2}{\omega^2}$ is about 3000 and the microwave should only marginally penetrate into the plasma. However, since $\frac{\nu_{en}}{\omega} = 1..4$, the collisions ensure that the microwave can still penetrate into the plasma. Since $\frac{\nu_{en}}{\omega}$ is about one, the energy transfer from the microwave to the plasma is optimal, as shown in section 1.2, and the plasma is heated well. This can also be seen when the refraction index is regarded: the imaginary part is $n_{Im} = 14.3$, which shows the good absorption of the microwave by the plasma. The real part is $n_{Re} = 16.9$ and contributes to a good absorption, since the phase velocity of the wave in the plasma is given by $v_{ph} = \frac{c}{n_{Re}}$, which shows that the phase velocity becomes smaller when the wave enters into the plasma. Since the phase velocity decreases, the microwave propagates slower and the power can be better absorbed by the plasma.

In section 1.1 different models to describe the plasma were presented. Since the electron density of atmospheric pressure microwave-generated plasma is in the range of $n_e \approx 10^{20}..10^{21}\,\text{m}^{-3}$, the APS plasmas would be best described by the collisional-radiative model. However, the description via the collisional-radiative model is very complex since rate equations for all populating and depopulation processes must be formulated. Therefore, for a first characterisation of the APS plasma a simpler model was chosen. Atmospheric pressure microwave-generated plasmas are collisionally dominated plasmas with an electron neutral particle collision frequency of $\nu_{en} \approx 2..7 \cdot 10^{10}\,\text{m}^{-3}$, which is considerably higher than the transition probability of $A_\nu^n = 2.45 \cdot 10^7\,\text{s}^{-1}$ of a commonly observed transition in nitrogen plasmas [49]. Thus, since $\frac{\nu_{en}}{A_\nu^n} \approx 8 \cdot 10^2..3 \cdot 10^3$, deexciting collisions

1.3. TYPICAL PLASMA PARAMETERS

Table 1.1: The table summarises the fundamental and derived plasma parameters of atmospheric pressure microwave-generated plasmas.

fundamental plasma parameters:	
gas temperature T_g	$T_g \approx 1500..4000\,\text{K}$
electron temperature T_e	$T_e \approx 4000..20000\,\text{K}$, which corresponds to an energy of $T_e \approx 0.345..1.726\,\text{eV}$
neutral particle density n_a	at a gas temperature of about $T_g \approx 3000\,\text{K}$ and at atmospheric pressure: $n_a \approx 2.1 \cdot 10^{24}\,\text{m}^{-3}$
electron density n_e	$n_e \approx 10^{20}..10^{21}\,\text{m}^{-3}$
electron neutral collision frequency ν_{en}	$\nu_{en} \approx 2..7 \cdot 10^{10}\,\text{s}^{-1}$, [27]
derived plasma parameters:	
degree of ionisation χ	$\chi \approx 10^{-3}..10^{-5}$
Debye length λ_D	$\lambda_D \approx 0.1..1\,\mu\text{m}$
plasma parameter N_D	number of particles in the Debye sphere: $N_D \approx 40..350$
plasma frequency ω_p	$\omega_p = 560 \cdot 10^9 ... 1800 \cdot 10^9\,\text{s}^{-1}$
cutoff density n_c	$n_c = 7.4 \cdot 10^{16}\,\text{m}^{-3}$
refraction index n_{ref}	$n_{Re} = 14.3;\quad n_{Im} = 16.9$
skin depth δ of the microwave	$\delta \approx 1.2\,\text{mm}$
absorbed power $\frac{p_w}{E^2}$	$\frac{p_w}{E^2} \approx 66.3\,\frac{\text{W}}{\text{m}^3(\text{V/m})^2}$

can not be neglected in the description of the APS plasma. Therefore, a description of the APS plasma by the corona model can be debarred since this model neglects collisional deexcitation.

A description by a partial local thermodynamic equilibrium seems to be more suitable for a first characterisation of the APS plasma. Here the excitation is given by the Boltzmann distribution and the ionisation by the Saha-Eggert equation. The radiation of the plasma depends on the particle densities and temperatures as well as on the complex quantum mechanical structure of the molecules and atoms. So the radiation of the plasma, which is voluntarily emitted by the plasma, can be used to acquire information about the particle densities and temperatures. The determination of temperatures and densities or generally the characterisation of plasmas by the voluntarily emitted radiation of the plasma is called optical emission spectroscopy (OES).

So a first characterisation of the APS plasma was performed by optical emission spectroscopy and it was assumed, that the APS plasma is in partial local thermodynamic equilibrium. However, for a detailed and comprehensive characterisation of the APS plasma a collisional-radiative model should be applied in further works.

Chapter 2

Development of the Atmospheric Microwave Plasma Source (APS)

In this chapter at first contemporary used atmospheric plasma sources and their deficiencies are presented. Thereof, the requirements of an atmospheric microwave plasma source for successful use in industrial application are drafted and the design and concept of the atmospheric microwave plasma source (APS) are introduced in section 2.1.

Thereafter, the detailed development of the APS is presented: For a successful industrial application a simple ignition of the plasma as well as stable plasma operation are indispensable. To guarantee an ignition without any additional igniters a high electric field must be reached and therefore detailed information about the electric fields is required which is why finite element simulations of the electric field distribution were conducted and are presented in section 2.3.

The simulation results were verified by measurements with a network analyser and are presented in section 2.4 and resulted in an APS configuration which provides plasma ignition without any additional igniters as well as stable plasma operation. This is presented in chapter 3.

2.1 Design of the Atmospheric Microwave Plasma Source APS

At present a variety of different configurations of atmospheric pressure microwave plasma sources at a frequency of 2.45 GHz exists [39]. Some of them are used to study the decomposition of waste gases [16, 17, 18, 19, 20, 21, 22, 23, 24]. Examples for atmospheric pressure microwave plasma sources are the surfatron [2, 3, 4], waveguide-based axial-type microwave plasma sources [5, 6, 7, 8, 9], (tapered) waveguide-based systems [10, 11, 12, 13, 14, 15] and resonator-based microwave plasma sources [16, 17, 18, 19]. Three different principles of atmospheric pressure microwave plasma sources will be presented in the following sections. First, two waveguide-based axial-type microwave plasma sources which were first proposed by M. Moisan, second, (tapered) wave-guide systems, and third, cylindrical resonator based sources will be described. Afterwards, the design of the atmospheric pressure microwave plasma source APS, which is enhanced and

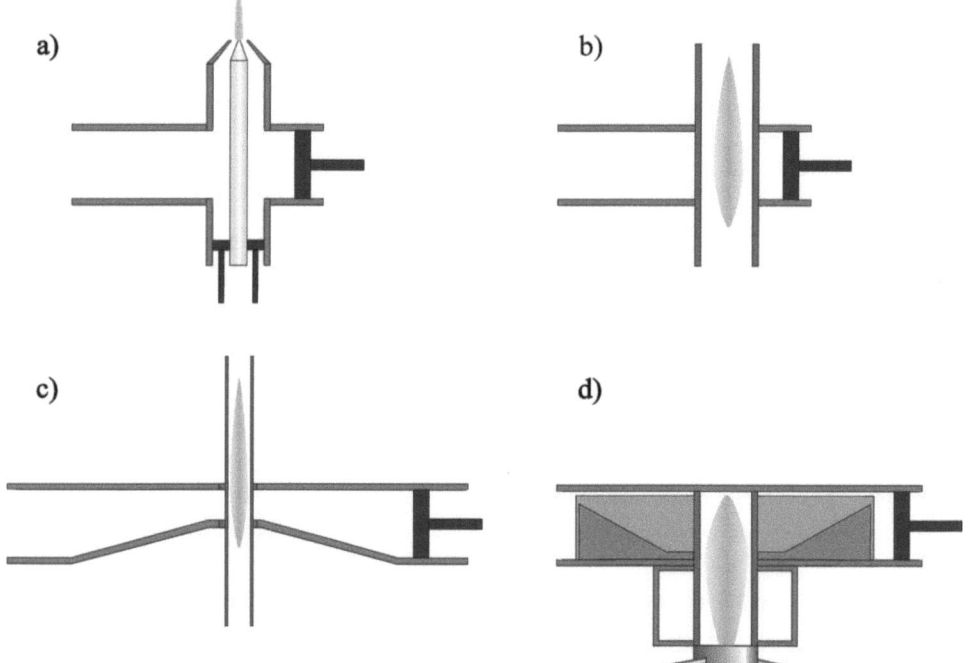

Figure 2.1: Different kinds of atmospheric pressure microwave plasma sources: a) waveguide-based axial-type microwave plasma source 'torch à injection axiale' (TIA), introduced by Moisan et al. [2, 3, 4], b) the waveguide-based system 'Microwave Plasma Torch' (MPT), developed by Jin et al. [40], c) another waveguide-based system: a surfaguide presented by M. Moisan et al. [25], and d) a cylindrical resonator based plasma source developed by Baeva et al. [16, 17, 18, 19].

characterised in this work, will be introduced and compared to the other sources.

The first waveguide-based axial-type microwave plasma source was introduced by M. Moisan and is called TIA, 'torch à injection axiale' [5]. Fig. 2.1a) shows a schematic view of the TIA design. The TIA is based on a conventional waveguide-to-coaxial line transition. The plasma gas is guided through the inner conductor of the coaxial line and exits through a nozzle. The microwave is guided in via the described rectangular and coaxial wave guides, reaches the end of the coaxial transition line and with the gas flow through the inner conductor a plasma is ignited at its top. Two shorting plungers are used to match the impedances and to minimise the reflected power. The plasma forms a small flame at the top of the inner conductor.

The simplest waveguide-based system is the Microwave Plasma Torch (MPT) and was first suggested and developed by Jin et al. [39, 40]. Its design is very simple and is shown in Fig. 2.1b). A quartz tube is inserted perpendicular in a rectangular waveguide and a longitudinal extended plasma can be ignited. Again, a shorting plunger is used to match the impedances and to minimise

2.1. DESIGN OF THE ATMOSPHERIC MICROWAVE PLASMA SOURCE APS

the reflected microwave power. The so called surfaguide has a similar design, which was developed by M. Moisan and is shown in Fig 2.1c) [25]. The surfaguide consists of several parts: first a common rectangular waveguide which is tapered to a part with a reduced height and then is again tapered to the common height. In the middle of the part with the reduced height, a thin quartz tube is inserted perpendicular into the waveguide, as can bee seen in Fig 2.1c). On one side of the waveguide the microwave is coupled in and guided to the quartz tube where a plasma can be ignited. Since the quartz tube is very thin a complex cooling of the tube is indispensable. On the other side of the waveguide a shorting plunger again is necessary to match the impedances and to minimise the reflected power.

Another contemporary atmospheric pressure microwave plasma source is based on a cylindrical resonator which is coupled to a rectangular waveguide as shown in Fig 2.1d) and was developed by Baeva et al. [16, 17, 18, 19]. The resonator has a very high quality and the coupling between the waveguide and the resonator is improved by a tapered coupling element. This allows a self ignition of the plasma but only a pulsed plasma operation is possible since the plasma detunes the resonant frequency of the resonator and because of the high quality no more microwave power can be coupled into the resonator.

All of the contempory atmospheric pressure microwave plasma sources presented above have some disadvantages and therefore they are inappropriate for a successful application in industrial processes such as the purification of waste gases. For example, the TIA only produces a very small plasma flame so that only little amounts of waste gases can be treated while the surfaguide needs a complex cooling system of the quartz tube. Other disadvantages are that additional igniters such as AC electric sparks or torch modules are often needed or that self igniting resonator based plasma sources can only operate in pulsed mode [16, 17, 18, 19, 20, 21, 22, 23, 24].

However, a plasma system which will be used in industrial applications has to be easy to handle and must be able to treat large amounts of gases. Therefore, a self igniting system, which can operate continuously and is able to treat large amounts of gases, is needed. A large diameter quartz tube allows the treatment of huge gas flows due to the long dwell time. A large enough diameter prevents contact with the plasma avoiding thermal damage to the quartz tube. On the other hand the diameter of the quartz tube must be smaller than half the wavelength of the microwave since otherwise microwave would be radiated. For an ignition of the plasma at atmospheric pressure in air and without any additional igniters a high electric field of $E_{ignition} \approx 2 \cdot 10^6 \frac{V}{m}$ inside of the plasma source is required. A high electric field can be obtained when resonators, which have a high quality Q, are used. However, such resonators have a narrow resonance curve with a sharp resonant frequency. This can lead to further problems: On the one hand the supplied microwave must have exactly the same frequency as the resonant frequency of the resonator, on the other hand the resonant frequency of the resonator shifts due to the permittivity of the burning plasma, the microwave can no longer penetrate into the resonator and no continuous operation of the plasma source is possible.

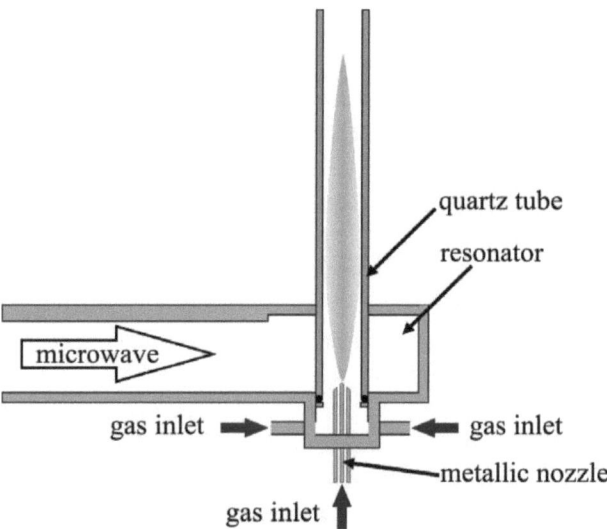

Figure 2.2: Schematic view of the atmospheric pressure microwave plasma source APS, enhanced and characterised in the present work: It is based on an axially symmetric cavity. The microwave is coupled in by a rectangular waveguide and the plasma is confined in a quartz tube which is located in the centre of the cavity.

Thus an optimal plasma source which provides plasma ignition without any additional igniters as well as stable and continuous plasma operation is provided by a configuration which has a high enough quality that a sufficiently high electric field for the ignition can be established but the quality must also be low enough that a continuous operation is possible.

The atmospheric pressure microwave plasma source APS, which was enhanced and characterised in this work, is based on an axially symmetric cavity which acts as a cylindrical resonator. Microwaves of a frequency of 2.45 GHz are fed into the cavity via rectangular wave guides. A quartz tube in the centre of the resonator confines the plasma and acts as a reaction chamber. The quartz tube has an inner diameter of 26 mm and an outer diameter of 30 mm and therefore is large enough for the treatment of large gas flows but also smaller than half the wavelength of the used microwave. A metallic nozzle which forms a coaxial part below the resonator is needed for the gas inlet. Since this gas inlet forms a coaxial structure below the cylindrical resonator a second resonator is created as will be seen in section 2.3.2.3 which describes finite element simulations of the electric field distribution of the resonator configurations. So this plasma source is actually based on two resonators. A second tangential gas inlet which leads to a swirl flow of the plasma gas increases stable plasma operation. Fig 2.2 shows a schematic view of the plasma source.

Thus to develop a self igniting plasma source, detailed knowledge about the electric field distribution in the resonator configurations is necessary. Simple configurations such as a sole cylindrical resonator can be calculated analytically as presented in the following section 2.2 but more complex systems like the complete plasma source with the quartz tube and the metallic nozzle can only be covered by using finite element simulations of the electric field distribution. This will be presented in section 2.3.

2.2 Analytical Solutions of the Maxwellian Equations for Cylindrical Resonators

Simple configurations such as a cylindrical resonator with a height h and a radius r can be solved analytically by solving Maxwell's equations. In a charge-free and insulating dielectric medium

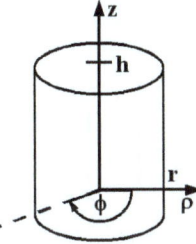

Figure 2.3: Cylindrical resonator with a height h and a radius r.

particularly in vacuum the density of volume charge ρ_e is zero and the specific conductivity σ is also zero. Therefore, Maxwell's equations for sinusoidal signals with a time dependence of $e^{-i\omega t}$ are given by [50]:

$$\nabla \cdot \vec{E} = 0 \tag{2.1a}$$

$$\nabla \cdot \vec{B} = 0 \tag{2.1b}$$

$$\nabla \times \vec{E} = i\frac{\omega}{c}\vec{B} \tag{2.1c}$$

$$\nabla \times \vec{B} = -i\mu\epsilon\frac{\omega}{c}\vec{E}, \tag{2.1d}$$

where $\vec{E}$ and $\vec{B}$ are the electric and magnetic field components, respectively, ω the angular frequency, c the speed of light in vacuum, μ the permeability of free space, and ϵ the permittivity of free space. Resolving (2.1c) to $\vec{B}$ and inserting in (2.1d) or resolving (2.1d) to $\vec{E}$ and inserting in (2.1c) one gets the two following wave equations [50]:

$$\nabla^2 \vec{E} + \frac{\omega^2}{c^2}\mu\epsilon\vec{E} = 0 \tag{2.2a}$$

$$\nabla^2 \vec{B} + \frac{\omega^2}{c^2}\mu\epsilon\vec{B} = 0. \tag{2.2b}$$

Since we have cylindrical symmetry, the $\vec{E}$- and $\vec{B}$-components can be written as $\vec{E} = E(x,y,z) \cdot e^{-i\omega t}$ and $\vec{B} = B(x,y,z) \cdot e^{-i\omega t}$ and the z-component can be separated:

$$\vec{E} = E(x,y) \cdot e^{-i\omega t \pm ikz} \tag{2.3a}$$

$$\vec{B} = B(x,y) \cdot e^{-i\omega t \pm ikz}. \tag{2.3b}$$

According to the boundary conditions for standing waves, the z-components are given by [50]:

$$E_z = E(x,y) \cdot e^{-i\omega t} \cdot \cos\left(\frac{l\pi z}{h}\right) \tag{2.4a}$$

$$B_z = B(x,y) \cdot e^{-i\omega t} \cdot \sin\left(\frac{l\pi z}{h}\right) \tag{2.4b}$$

where l is the axial wave number which can be $l = 0, 1, 2, 3, \ldots$ for the electric field and $l = 1, 2, 3, \ldots$ for the magnetic field component. Changing to cylindrical coordinates ρ, ϕ, and z and formulating a separation ansatz due to the rotational symmetry one gets [50]:

$$\left.\begin{array}{c} E(\rho, \phi) \cdot e^{\pm ikz - i\omega t} \\ B(\rho, \phi) \cdot e^{\pm ikz - i\omega t} \end{array}\right\} = \left\{\begin{array}{c} \Phi(\rho) \cdot e^{im\phi} \cdot e^{\pm ikz - i\omega t} \\ \Psi(\rho) \cdot e^{im\phi} \cdot e^{\pm ikz - i\omega t}, \end{array}\right. \tag{2.5}$$

and thus the two dimensional wave equations can be written in the following form [50]:

$$\left(\frac{\partial^2}{\partial \rho^2} + \frac{1}{\rho}\frac{\partial}{\partial \rho} + \gamma^2 - \frac{m^2}{\rho^2}\right)\Psi(\rho) = 0 \tag{2.6}$$

which is the so called Bessel differential equation with [50]

$$\gamma^2 = \mu\epsilon\frac{\omega^2}{c^2} - k^2, \quad \text{where} \quad k = \frac{l\pi}{h}. \tag{2.7}$$

The solution of the Bessel differential equation is given by the Bessel function J_{mn}. Herefrom the field components can be calculated, the z-component of the electric field, for example, is given by [50]:

$$E_z(\rho, \phi, z, t) = E_0 \cdot e^{-i\omega t} \cos\left(\frac{l\pi z}{h}\right) J_{mn}(\gamma_{mn} \cdot \rho) \cdot e^{im\phi} \tag{2.8}$$

with m and n being radial and azimuthal wave numbers, respectively. m and n take on the values $m = 0, 1, 2, \ldots$ and $n = 1, 2, 3, \ldots$. The boundary condition $E_z(\rho = r) = 0$ results in:

$$\gamma_{mn} = \frac{x_{mn}}{r} \tag{2.9}$$

where x_{mn} is the nth-root of the equation $J_m(x) = 0$. Combining 2.7 and 2.9 the resonant frequency results in ω_{mnl}:

$$\omega_{mnl} = \frac{1}{r\sqrt{\epsilon\mu}}\sqrt{x_{mn}^2 + \left(\frac{l\pi r}{h}\right)^2}. \tag{2.10}$$

2.3. SIMULATIONS WITH COMSOL MULTIPHYSICS^{TM}

The resonator of the atmospheric pressure microwave plasma source APS has a height h of 0.0482 m and a radius r of 0.05 m. So the resulting resonant frequency of the lowest E-mode with $m = 0, n = 1$, and $l = 0$, the E_{010}-mode, is

$$\omega_{010} = \frac{x_{01}}{r\sqrt{\epsilon\mu}} \qquad (2.11)$$

and with $x_{01} = 2.405$

$$\nu_{010} = \frac{\omega_{010}}{2\pi} = 2.296\,\text{GHz}. \qquad (2.12)$$

Since the magnetron's frequency is fixed at 2.45 GHz a relation to calculate the dimensions of the resonator for a fixed frequency is required. Considering only E_{mn0}-modes 2.10 is only dependent on the resonator radius and can be resolved as follows:

$$r_{mn0} = \frac{x_{mn}}{\omega\sqrt{\epsilon\mu}} \qquad (2.13)$$

By now the appropriate resonator radius can easily be calculated for a given magnetron frequency. The magnetron frequency is fixed at 2.45 GHz which results in a resonator radius of $r = 0.0468$ m. Since the radius of the APS is 0.05 m, which results in a resonant frequency of $\nu_{010} = 2.296$ GHz for the E_{010}-mode, the radius of the APS is too large to be resonant to the magnetron's frequency of 2.45 GHz if it is only regarded as a cylindrical resonator.

However, the real resonator configuration is more complex as described in section 2.1. To confine the plasma, a quartz tube in the centre of the resonator is necessary. Furthermore, a metallic nozzle is needed for the gas inlet which forms a coaxial part below the resonant cavity. These complex configurations cannot be calculated analytically anymore. Therefore, simulations of the electric field distribution of these complex configurations are indispensable. The commercial finite element simulation software COMSOL Multiphysics^{TM} was chosen for this purpose. The following section 2.3 describes the simulation software COMSOL Multiphysics^{TM} in section 2.3.1 while section 2.3.2 describes the simulation results.

2.3 Simulations with COMSOL Multiphysics^{TM}

2.3.1 The Simulation Software COMSOL Multiphysics^{TM}

COMSOL Multiphysics^{TM} is a commercial finite element simulation software which can be used to solve three dimensional scientific problems based on partial differential equations [51]. It has different kinds of physics modes which consist of predefined templates and user interfaces. These are already equipped with the required equations and variables so that only the geometry of the explorative configurations and the relevant physical quantities such as material properties - here especially the conductivity - must be defined rather than all the underlying equations. COMSOL Multiphysics^{TM} then compiles a set of essential partial differential equations which describe the whole model. The different physics modes can be used isolated or they can be combined to study multiphysical phenomena. Here only a single physics mode, the RF module, was used, which is configured for the analysis of electromagnetic waves and will be described in the following section 2.3.1.1. COMSOL Multiphysics™ has three different ways of providing the partial differential equations. They are described by the following three mathematical application modes:

- coefficient for linear or almost linear models,

- general form for non-linear, and

- weak form for models with partial differential equations on, for example, boundaries, edges or for models which have mixed time and space derivatives.

With these application modes, the following analysis types can be made:

- stationary and time-dependent analysis,

- linear and non-linear analysis, and

- Eigenfrequency and modal analysis.

As already mentioned above, the partial differential equations are solved by the finite element method. Additionally, the meshing is adapted and the errors are controlled by using different numerical solvers which were usually chosen automatically and adapted to the problem by COMSOL Multiphysics™ itself.

2.3.1.1 The RF Module

The RF Module is designed to solve and model scientific problems concerning the propagation of electromagnetic waves. This module can handle time-harmonic, time-dependent, as well as Eigenfrequency and Eigenmode tasks. For the propagation of electromagnetic waves this means that harmonic, transient, and Eigenfrequency and Eigenmode analyses are possible. The module has two application modes for three dimensional problems: the electric waves application, which was chosen, and the boundary mode analysis application. The electromagnetic waves application provides, among others, harmonic propagation and Eigenfrequency analysis. With the harmonic propagation, stationary or steady state problems can be solved. It was therefore used to simulate the steady state electric field distribution for a given configuration when a microwave is coupled into the configuration. The Eigenfrequency analysis was taken to calculate the Eigenfrequencies as well as the electric field distribution of a given configuration.

The Maxwellian equations form the basis [51]. The limited conductivity which gives rise to Ohmic losses is realised as a complex permittivity $\epsilon_r = \epsilon' - j\epsilon''$. ϵ'' reflects all the losses whereas ϵ' is

2.3. SIMULATIONS WITH COMSOL MULTIPHYSICS™

the real part of ϵ_r. Alternatively, the losses can be expressed by the conductivity σ itself using $\sigma = \omega \epsilon''$ where ω is the angular frequency. The partial differential equation for time-harmonic and Eigenfrequency problems can then be written in the following form:

$$\nabla \times (\mu^{-1} \nabla \times \vec{E}) - k_0^2 \epsilon \vec{E} = 0 \tag{2.14a}$$

$$\nabla \times (\epsilon^{-1} \nabla \times \vec{H}) - k_0^2 \mu \vec{H} = 0 \tag{2.14b}$$

[51], with $\vec{D} = \epsilon \vec{E}$ and $\vec{B} = \mu \vec{H}$, μ, ϵ are the permeability and permittivity, respectively, and $k_0 = \omega \sqrt{\epsilon_0 \mu_0} = \omega/c$ (c: speed of light in vacuum) being the wave number of free space ($\epsilon = \epsilon_r \epsilon_0$, $\mu = \mu_0 \mu_r$). When Eigenfrequency problems are solved the Eigenvalue is $\lambda = k_0^2$.

The boundary and interface conditions can be written as follows:

$$\vec{n}_2 \times (\vec{E}_1 - \vec{E}_2) = 0 \tag{2.15a}$$

$$\vec{n}_2 \cdot (\vec{D}_1 - \vec{D}_2) = \rho_s \tag{2.15b}$$

$$\vec{n}_2 \times (\vec{H}_1 - \vec{H}_2) = \vec{J}_s \tag{2.15c}$$

$$\vec{n}_2 \cdot (\vec{B}_1 - \vec{B}_2) = 0 \tag{2.15d}$$

[51], where $\vec{n}$ is the normal vector, ρ_s, and $\vec{J}_s$ are the surface charge density and the surface current density, respectively. Almost all simulations were performed with perfect electric conductor boundaries since the simulations with real conductivities led to nearly the same results for the electric field distributions. When perfect electric conductors are used as boundaries, equation 2.15 is reduced to:

$$\vec{n} \times \vec{E} = 0 \tag{2.16}$$

[51]. The electric field distribution was calculated by using harmonic propagation. Therefore, a microwave has to be induced in the created configuration. This microwave excitation is generated by a matched boundary condition at the front of the waveguide and the boundary conditions become:

$$\vec{n} \times (\nabla \times \vec{E}) - j\beta(\vec{E} - (\vec{n} \cdot \vec{E})\vec{n}) = -2j\beta(\vec{E}_0 - (\vec{n} \cdot \vec{E}_0)\vec{n}) \tag{2.17}$$

where $\vec{E}$ is the independent variable and $\vec{E}_0$ is the incident field or the source field [51]. β is the propagation constant and for the propagation of an E_{01}-mode in the used rectangular waveguide given by:

$$\beta = \sqrt{(2\pi\nu)^2 \epsilon \mu - \left(\frac{\pi}{a}\right)^2} \tag{2.18}$$

with ν being the frequency of the microwave and a being the longer side of the rectangular waveguide. So the excitation is given by an incident electric field and the power can be varied by the amplitude E_0 of a sinusoidal electric field. Before the problem can be solved a mesh must be created which consists of tetrahedral elements for three-dimensional configurations. A custom mesh size of $\leq 1/10\lambda$ was used for the simulations where λ is the wavelength of the microwave. In narrow regions the mesh size was refined. The problems are solved with different solvers which are usually chosen automatically and adapted to the problem. Normally, the stationary linear solver is used for harmonic propagation and the Eigenvalue solver is chosen for the Eigenfrequency analysis [51]. Some simulations were performed where a parameter was varied for example the frequency of the incident microwave. Therefore, harmonic propagation was used and the parametric linear solver was selected as solver [51]. Many different physical values are accessible for the evaluation of the result. Here the interesting physical values are the electric field components and their norms, the absorbed microwave power in particular parts of the configuration, the incident and reflected power as well as the resonant frequencies. The electric field components can be displayed as slices at different places in the configuration and at different intersection lines.

So the general procedure of all simulations performed with COMSOL Multiphysics^{TM} can be summarised as follows:

1. The geometry of the whole configuration is created.

2. The properties of individual parts, such as the electrical properties of the quartz tube, of the configuration are assigned.

3. The boundary conditions are defined, such as perfect electric conductors for the metallic walls and the matched boundary condition at the front of the waveguide where the microwave is excited.

4. The application mode and the solver are chosen: harmonic propagation or Eigenfrequency analysis

5. The precision of the mesh is defined.

6. The simulation is run, the results are displayed and evaluated.

The simulations performed with COMSOL Multiphysics^{TM} and their results are presented and discussed in the following section 2.3.2.

2.3.2 Simulation Results

For a successful operation of the atmospheric pressure microwave plasma source APS in industrial processes, a self igniting APS which needs no additional igniters is indispensable. Therefore, knowledge about the electric field distribution of different APS configurations including the quartz tube and the metallic nozzle for the gas inlet is needed. Simple configurations such as a cylindrical resonator can be calculated analytically as shown in section 2.2. For more complex configurations like the realistic APS with the quartz tube and the coaxial part below, formed by the gas inlet, the electric field distribution of the configuration is only accessible by using simulations. These

2.3. SIMULATIONS WITH COMSOL MULTIPHYSICS^{TM}

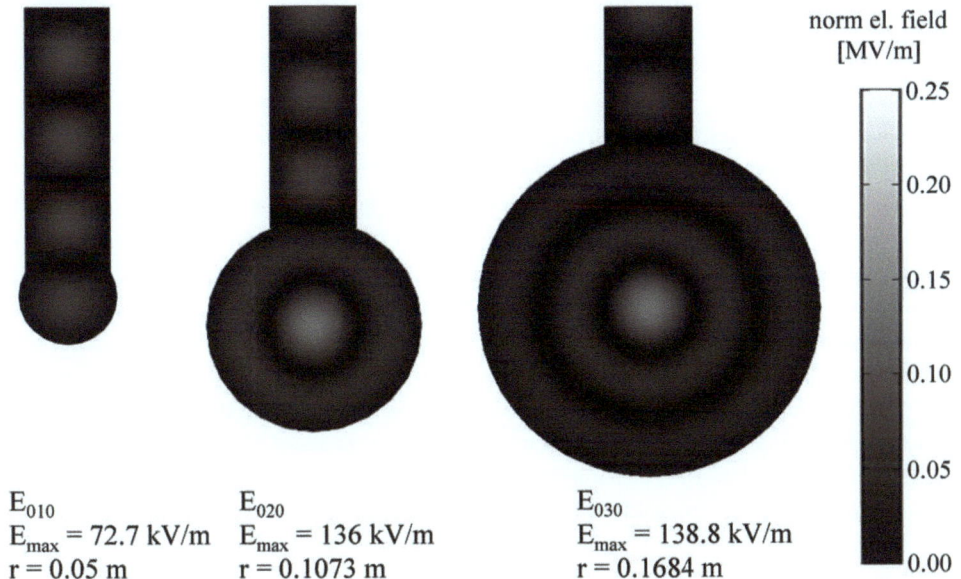

E_{010}
E_{max} = 72.7 kV/m
r = 0.05 m

E_{020}
E_{max} = 136 kV/m
r = 0.1073 m

E_{030}
E_{max} = 138.8 kV/m
r = 0.1684 m

Figure 2.4: View from above on the distribution of the norm of the electric field of a) a cylindrical resonator with the same radius as the APS, $r = 0.05$ m, b) a E_{020}-mode resonator with a radius of $r = 0.1073$ m, and c) a E_{030}-mode resonator with a radius of $r = 0.1684$ m. The excitation of the microwave was carried out at the front of the waveguide with a frequency of 2.45 GHz and a power of 3 kW.

simulations were performed using the simulations software COMSOL Multiphysics^{TM} which is described in the previous section 2.3.1. In this section 2.3.2 the results of the simulations are presented and discussed. The simulations were performed using the procedure described in section 2.3.1.1.

2.3.2.1 Simulations of Cylindrical Resonators Excited in Higher Modes

First, Eigenfrequency analysis of a cylindrical resonator with the radius of the atmospheric pressure microwave plasma source APS of $r = 0.05$ m and of a E_{010}-mode resonator for 2.45 GHz ($r = 0.0468$ m) were performed. Resonant frequencies of 2.296 GHz and 2.45 GHz, respectively, resulted, which are in excellent agreement with the analytical calculations performed in section 2.2. Then the electric field distribution of a cylindrical resonator with a radius of the APS ($r = 0.05$ m) coupled to a rectangular waveguide was simulated. The microwave was excited at the front of the waveguide with a power of 3 kW and a frequency of 2.45 GHz. Fig. 2.4a) shows the norm of the electric field. It can easily be seen, that the maximal electric field of about $E = 72.7$ kV/m in the centre of the resonator is not higher compared to the one of about $E = 82.8$ kV/m in the waveguide. This is not surprising since the resonator diameter of $r = 0.1$ m has nearly the same size as the waveguide width ($a = 0.0864$ m and $b = 0.0432$ m) as can be seen in Fig. 2.4a). So the E_{010}-mode resonator is only a rounded short-circuited waveguide. The small electric field in the centre of the resonator reflects the poor quality Q of the configuration consisting of the resonator and the waveguide.

The quality Q describes the relation of the energy stored in the resonator per time to the energy loss. The losses are caused by Ohmic losses in the metallic resonator walls on the one hand. On the other hand, microwave power, which is reflected at the resonator walls, is able to leave the resonator through the waveguide. A high electric field in the resonator centre can be achieved by increasing the power of the microwave. However, thereby the electric field in the waveguide is increased, too. If only the electric field in the resonator centre shall be increased, the quality of the resonator must be improved. The insufficient quality caused by Ohmic losses can be easily improved, if the resonator walls are coated with a thin metal coating, which has a high conductivity. The other loss channel caused by the loss of the reflected microwave through the waveguide is much profounder. Since the microwave is coupled into the resonator via a waveguide with nearly the same dimensions as the resonator diameter, a remarkable amount of the microwave power, which is reflected at the resonator walls, can leave the resonator through the waveguide without any difficulties.

Since an electric field of about 2 MV/m is required for the ignition of plasma in air at atmospheric pressure, a sole E_{010}-mode resonator cannot be used for a self igniting plasma source at atmospheric pressure, if only microwave sources of some kW are available. Therefore, a configuration with a higher quality, which leads to a higher electric field in the resonator centre, is required. An improvement of the quality can be achieved, when the resonator diameter is enlarged compared to the waveguide dimensions, so that less reflected microwave power is lost. This means that an E_{010}-mode resonator can no longer be used, instead, we have to pass on to higher mode resonators.

When only E_{lm0}-mode resonators are considered, the radius for such a cylindrical resonator can be calculated by formula 2.13

$$r_{mn0} = \frac{x_{mn}}{\omega \sqrt{\epsilon \mu}}$$

given in section 2.2. The electric field distributions of higher mode resonators were analysed to examine if configurations with higher mode resonators lead to configurations with a sufficiently high quality and thus to a sufficiently high electric field for the ignition of plasma. According to equation 2.13 the radius of an E_{020}-mode resonator is $r_{020} = 0.1073$ m. Fig. 2.4b) shows the electric field distribution of an E_{020}-mode resonator coupled to a rectangular waveguide. The resonator was again excited with a microwave of a frequency of 2.45 GHz and a power of 3 kW. The electric field in the centre of the E_{020}-mode resonator is increased and reaches a maximum of about $E = 136.3$ kV/m, which is noticeably higher compared to that in the centre of the E_{010}-mode resonator and to that in the waveguide of about $E = 82.5$ kV/m. This shows that the quality of the E_{020}-mode resonator has increased compared to the quality of the E_{010}-mode resonator. Going another step further, one comes to an E_{030}-mode resonator. The electric field distribution of an E_{030}-mode resonator coupled to a rectangular waveguide is shown in Fig. 2.4c). Again the quality of the resonator has improved which is reflected in the higher electric field of about $E = 138.8\,\text{kV/m}$ in the centre of the E_{030}-mode resonator. Further E_{0m0}-mode resonators show slightly higher electric fields but also have huge dimensions which are impractical to handle and are therefore not pursued any further.

These simulations revealed that higher mode resonators, which have larger diameters and therefore are much larger compared to the waveguide, have improved qualities Q, which is reflected in the higher electric field in their centre. Another way to improve the quality of the configuration can be achieved, if special coupling elements are used, which reduce the coupling hole between the waveguide and the resonator, so that less reflected microwave power is lost through this coupling hole. Therefore, different kinds of coupling elements were developed and explored which are presented in the following section 2.3.2.2.

2.3.2.2 Simulations of Configurations with Coupling Elements

As already shown in section 2.3.2.1, the quality, and accordingly the electric field in the centre of the resonator, can be improved when higher mode resonators are used. Since the dimensions of higher mode resonators are larger than the waveguide, less of the reflected microwave is lost through the waveguide, which corresponds to an enhanced coupling from the waveguide to the resonator. An improved coupling can also be achieved, if special coupling elements are used. These coupling elements reduce the dimension of the part of the waveguide, which is coupled to the resonator, so that the coupling hole between the waveguide and the resonator is small compared to the resonator itself.

Therefore, a waveguide with a reduced height and a length of $\lambda_h/4$, a so called $\lambda/4$-part, can be used. λ_h is the wavelength of the microwave in the waveguide and is given by the following equation:

$$\lambda_h = \frac{\lambda_0}{\sqrt{1 - \left(\frac{\lambda_0}{\lambda_c}\right)^2}}, \quad \text{here} \quad \lambda_h = 0.1728\,\text{m} \tag{2.19}$$

whereby λ_0 is the wavelength in free space and λ_c the critical wavelength of the waveguide given by [55]:

$$\lambda_c = \frac{1}{\sqrt{\left(\frac{m}{2a}\right)^2 + \left(\frac{n}{2b}\right)^2}} \tag{2.20}$$

with m and n being the mode numbers, here $m = 1$ and $n = 0$ and a and b the dimensions of the waveguide ($a = 0.864\,\text{m}$ and $b = 0.432\,\text{m}$). The length of $\lambda_h/4$ is chosen because reflections caused at the joint of the different parts superpose destructively. Fig. 2.5a) shows the electric field distribution of a cylindrical resonator with a radius $r = 0.05\,\text{m}$ (radius of the APS) coupled to a rectangular waveguide via a $\lambda/4$-part with a height of $h_{\lambda_h/4} = 0.005\,\text{m}$. A microwave with a frequency of 2.296 GHz, the resonant frequency of the cylindrical resonator, and a power of 3 kW was excited at the front of the waveguide. The electric field in the centre reaches a

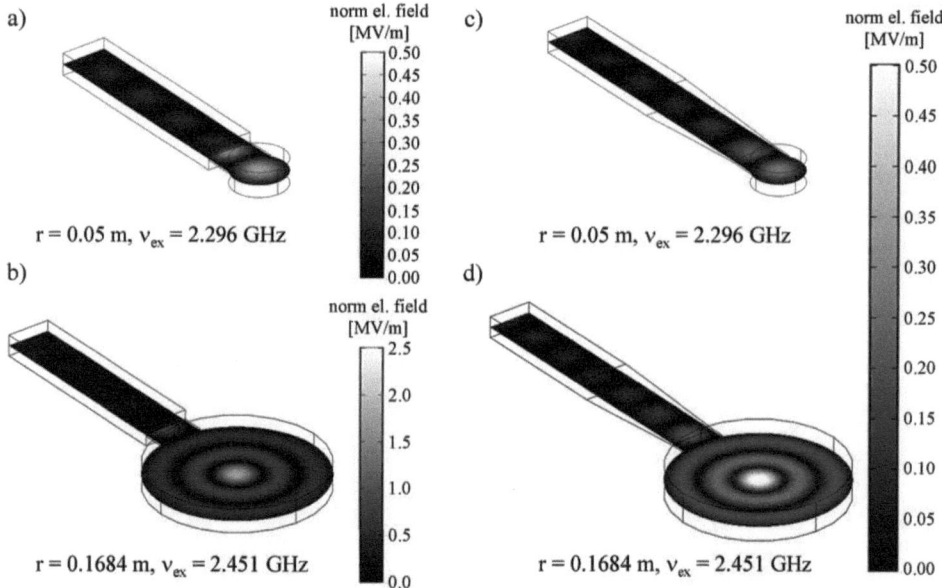

Figure 2.5: Distribution of the norm of the electric field of a) a cylindrical resonator with the same radius as the APS, $r = 0.05\,\text{m}$, coupled to the waveguide via a $\lambda/4$-part and c) via a taper, b) a E_{030}-mode resonator with a radius of $r = 0.1684\,\text{m}$ coupled to the waveguide via a $\lambda/4$-part and d) via a taper. The excitation of the microwave was carried out at the front of the waveguide with a frequency of $2.296\,\text{GHz}$ and $2.451\,\text{GHz}$ for the $r = 0.05\,\text{m}$ resonator and the E_{030}-mode resonator, respectively, and a power of $3\,\text{kW}$.

value which is $E = 0.36\,\text{MV/m}$ and is clearly higher than the electric field in the waveguide of $E = 0.09\,\text{MV/m}$. The coupling from the waveguide is improved since the coupling hole to the resonator is much smaller compared to the dimensions of the resonator. An even better improvement can be achieved if an E_{030}-mode resonator is coupled to the waveguide via a $\lambda/4$-part. The electric field distribution of such a configuration is shown in Fig. 2.5b). The electric field in the centre of the resonator reaches a maximum value of $E = 1.84\,\text{MV/m}$, which is much higher compared to that in the waveguide. Here the effect of the improvement of the quality is higher because the E_{030}-mode resonator itself has a larger dimension compared to the size of the waveguide and additionally the coupling hole between the waveguide and the resonator is reduced as well.

A softer transition from the waveguide with normal height to the resonator, with a reduced height, can be realised, when the hight of the waveguide is tapered along a distance of $n \cdot \lambda_h/4$. As a good compromise between an unmanageable length and a soft transition, a length of the tapered section of $l = 1.5\lambda_h$ has emerged. Fig. 2.5c) shows the electric field distribution of a waveguide coupled to a cylindrical resonator with a radius of $r = 0.05\,\text{m}$ (same radius as the APS) via such a taper. The excitation of the microwave was again carried out at the front of the waveguide with

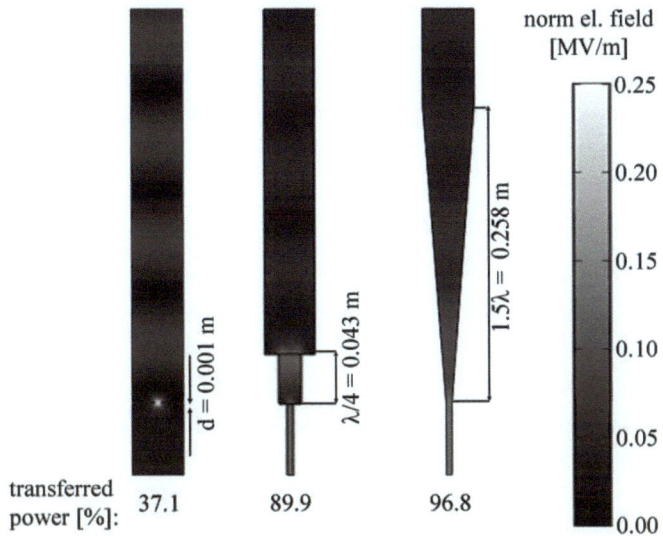

Figure 2.6: Transfer properties of the coupling elements: a) a slit, b) a λ/4-part, and c) a taper. The excitation of a microwave of a frequency of 2.45 GHz and a power of 3 kW was carried out at the front of the waveguide. The other side of the configurations was set as matched boundary with a propagation constant β for the incident wave.

a frequency of 2.296 GHz (resonant frequency of the $r = 0.05$ m resonator) and a power of 3 kW. The electric field of $E = 0.29$ MV/m in the centre of the resonator is a little bit smaller compared to the electric field, which was achieved with the configuration where the λ/4-part was used, but is distinctly higher than the electric field in the waveguide or in the case when no coupling element is used. When a E_{030}-mode resonator is used instead of the E_{010}-mode resonator, the electric field in its centre again can be increased to a value of $E = 0.58$ MV/m. The electric field distribution of this configuration is shown in Fig. 2.5d).

Furthermore, the transmission properties of these coupling elements were analysed, since for a stable and efficient plasma operation the reflected powers are required to be minimal. Additionally to the two coupling elements, the λ/4-part and the taper, a slit with a thickness of 0.001 m and height of 0.005 m was analysed. Fig. 2.6 shows the electric field distribution in the coupling elements if a microwave with a frequency of 2.45 GHz and a power of 3 kW is excited at the front of the waveguide. The other sides of the configurations were set as matched boundaries with a propagation constant β for the incident microwave. The particular distribution of the electric field is caused by reflections of the microwave at the different joints. The transmitted power ratio was calculated from the ratio of the incident microwave power to the transmitted power. The analysed slit has a transmitted power ratio of 37 % and has a high electric field in the gap. So this coupling element has poor transmission properties and additionally, the disadvantage, that already a sparkover at the gap could ignite a plasma, so that a plasma can no longer be ignited in the centre of the resonator and no stable plasma operation is possible. The transmitted power ratio through the $\lambda/4$-part reaches a much higher value of 89.9 %. The best transmission properties are provided by the taper with a transmitted power ratio of 96.8 %, since it forms a soft transition from the waveguide to the resonator.

These simulations of the electric field distribution of configurations with coupling elements and higher mode resonators showed that the quality, and, in accordance to that, the electric field in the resonator centre can be augmented significantly by using higher mode resonators and/or coupling elements. This offers the chance to design a self igniting plasma source. The quality of the configuration can be increased when the dimension of the resonator is significantly higher compared to the microwave supply line, which can be achieved by two different ways: Either to enlarge the resonator itself, which leads to higher mode resonators, or to reduce the coupling hole by using special coupling elements. The highest electric fields can be achieved, when both methods are combined, which results in an E_{030}-mode resonator in combination with a $\lambda/4$-part. However, for a stable and efficient plasma operation a high transmitted microwave power is desirable. Therefore, a configuration of a combination of a E_{030}-mode resonator with a taper, which provides the best transmission properties with 96.8 % transmitted microwave power and also has a relatively high electric field in the resoantor centre, is more preferable.

However, it must be noted, that the quality Q is linked to the resonant frequency ω by the following equation:

$$Q = \frac{\omega}{\Delta\omega}, \qquad (2.21)$$

with $\Delta\omega$ being the bandwidth. This means, that for a fixed frequency an improvement of the quality leads to a narrower resonant curve. A resonator configuration with a very high quality in turn has a sharp resonant frequency, which may cause problems in operation. Therefore, the frequency of the microwave generated by the magnetron needs to be very close to the resonant frequency. This can cause problems in operation since the magnetron frequency varies slightly in dependence of the output power and magnetron type. A burning plasma acts as dielectric causing a shift of the resonant frequency which results in further difficulties. Thus the practical implications have to be analysed which is presented in chapter 3. To examine if the quality of the improved configurations, consisting of a higher mode resonator and/ or combined with coupling elements, is sufficient for self ignition of a plasma and also suitable for stable and continuous plasma operation, an E_{030}-mode resonator, a taper, a $\lambda/4$-part, and a slit were designed and manufactured. The microwave properties of these configurations are analysed in section 2.4. The capability of plasma operations is tested and presented in chapter 3.

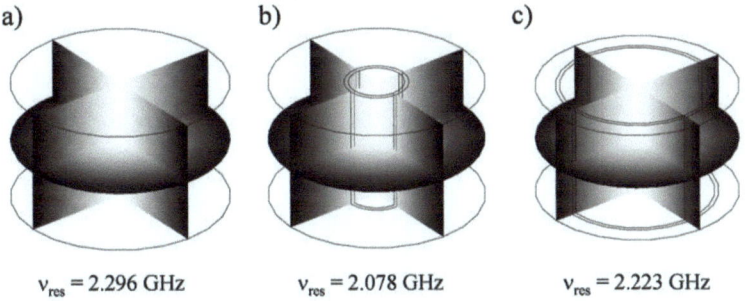

Figure 2.7: Eigenfrequency analyses of a cylindrical resonator with a radius of $r = 0.05\,\text{m}$ (same as APS): a) without a quartz tube, with an inserted quartz tube with b) outer diameter of 0.03 m and inner diameter of 0.026 m, and c) outer diameter of 0.08 m and inner diameter of 0.076 m.

2.3.2.3 Simulation of the Realistic APS

In section 2.3.2.1 simulations of the electric field distribution of a cylindrical resonator with the same radius as the APS of $r = 0.05\,\text{m}$ coupled to a waveguide were presented. However, the real APS has a quartz tube to confine the plasma and a metallic nozzle, which forms a coaxial structure below the cylindrical resonator. The influences of the quartz tube and the metallic nozzle will be analysed in this section using Eigenfrequency analyses as well as simulations of the electric field distribution of these configurations.

At first, the influence of the quartz tube, which is commonly used as reaction chamber in the experiments, on the resonant frequency was analysed. Eigenfrequency analyses of the cylindrical resonator (radius of $r = 0.05\,\text{m}$, same radius as the APS) with an inserted quartz tube were performed for this purpose. In Fig. 2.7 Eigenfrequency analyses of this cylindrical resonator with and without an inserted quartz tube are shown. The resonant frequency of the cylindrical resonator without the quartz tube is 2.296 GHz, as already mentioned in section 2.3.2.1. The Eigenfrequency analysis of the cylindrical resonator with an inserted quartz tube with an outer diameter of 0.03 m and an inner diameter 0.026 m revealed a resonant frequency of 2.078 GHz for this configuration. So the resonant frequency decreases when a quartz tube is inserted into the cylindrical resonator. This can be explained by the higher permittivity of the quartz tube compared to the permittivity of air. Thus the resonator is virtually enlarged, which leads to a virtually larger radius and results in a lower resonant frequency. In Fig. 2.7c) an Eigenfrequency analysis of a cylindrical resonator with an inserted quartz tube with other dimensions (inner diameter: 0.08 m, outer diameter: 0.076 m) is shown. It can be seen that this configuration has a resonant frequency of 2.223 GHz. This shows that the position of the dielectric is also important. It illustrates, that if the dielectric is located in a minimal electric field the shift of the resonant frequency is small but in contrast if it is located in a field maximum the shift is noticeably higher. So the inserted quartz tube causes a shift of the resonant frequency to lower frequencies and must be considered when an enhanced self

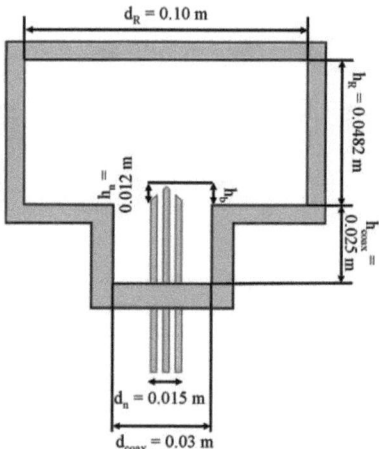

Figure 2.8: Dimensions of the APS with the metallic nozzle for the simulation of the electric field distributions and Eigenfrequency analyses: cylindrical resonator diameter $d_R = 0.1$ m, cylindrical resonator height $h_R = 0.0482$ m, diameter of the coaxial part below the cylindrical resonator $d_{coax} = 0.03$ m, height of the coaxial part $h_{coax} = 0.025$ m, height of the nozzle tip $h_n = 0.012$ m, diameter of the nozzle $d_n = 0.015\,4$, and height of the nozzle tip above the resonator base h_b.

igniting plasma source is developed. The effect that the inserted quartz tube virtually enlarges the resonator can be generalised for all other dielectrics but the magnitude of shift is characteristic to the dielectric material.

Furthermore, the influence of the metallic nozzle, - necessary for the gas inlet - which forms a coaxial structure below the cylindrical resonator, is explored. Simulations of the electric field distribution and Eigenfrequency analyses of the APS with the metallic nozzle were performed for this purpose. The dimensions of this APS are shown in Fig. 2.8.
Now the influence of the metallic nozzle position on the electric field distribution and resonant frequency is analysed. In the lower part of Fig. 2.9 the electric field distributions of three selected configurations are shown. The excitation of the microwave was carried out at the front of the waveguide with a frequency of 2.45 GHz and a power of 3 kW. The diagram shows the dependence of the resonant frequency and the norm of the electric field at the nozzle tip on the nozzle position. The simulations show, that the position of the nozzle has a distinct influence on the electric field distribution as well as on the value of the electric field. The left simulation shows the electric field distribution for a configuration where the nozzle extends 2 mm into the cylindrical resonator. The electric field reaches its maximum value of $E = 0.213$ MV/m at the nozzle tip. In the middle, a simulation of the configuration where the nozzle extends 8 mm into the resonator is shown. Again, the highest electric field is attained at the nozzle tip but its value of $E = 1.19$ MV/m has increased dramatically. The distribution of the electric field in the right simulation, where the nozzle ex-

2.3. SIMULATIONS WITH COMSOL MULTIPHYSICS^{TM}

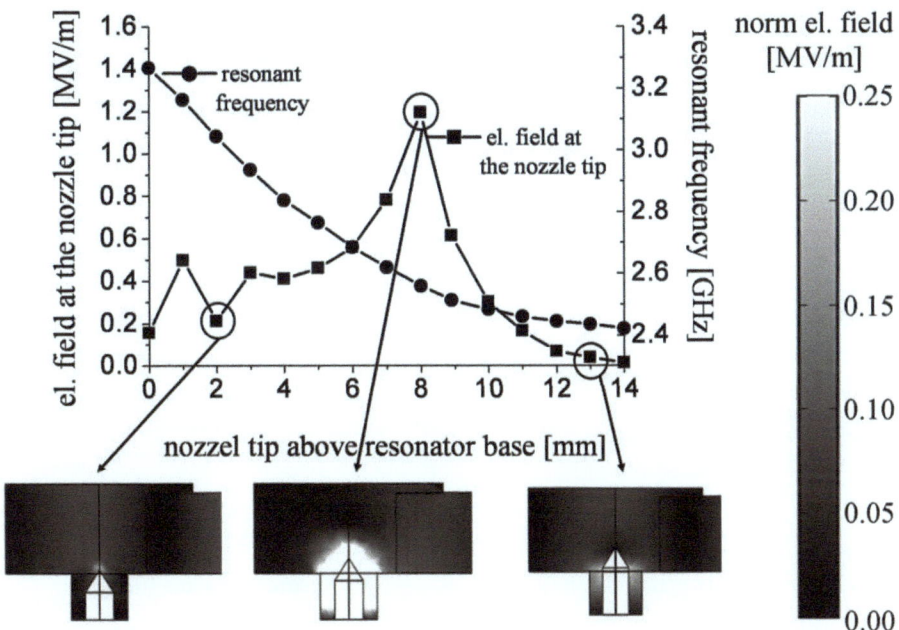

Figure 2.9: Simulation of the electric field distribution and Eigenfrequency analyses in dependence of the nozzle position: The diagram shows the norm of the electric field at the nozzle tip and the resonant frequency in dependence of the nozzle position. Below, the electric field distribution of three selected configurations is shown. The norm of the electric field reaches its highest value when the metallic nozzle extends 8 mm into the cylindrical resonator. The resonant frequency decreases when the nozzle extends further into the cylindrical resonator.

tends 13 mm into the resonator, shows that the situation has changed. The highest electric field of $E = 0.271$ MV/m is no longer reached at the nozzle tip ($E = 0.038$ MV/m) but at the base of the cylindrical resonator.

The dependence of the electric field at the nozzle tip from the nozzle position is summarised in the diagram in Fig. 2.9, which shows, that the highest electric field can be achieved when the nozzle extends 8 mm into the cylindrical resonator. The circles in the diagram show the resonant frequency in dependence of the nozzle position. It shows that the resonant frequency decreases from 3.266 GHz when the nozzle extends 0 mm into the resonator to 2.421 GHz when it extends 14 mm into the resonator. So the resonant frequency decreases when the nozzle is moved into the resonator. This means on the other hand, that the nozzle can be used to adjust the resonant frequency and therefore can be used as a tuning element.

Thus these simulations revealed how dramatically the metallic nozzle affects the resonant frequency as well as the distribution and the value of the electric field. A maximum electric field can be achieved when the nozzle extends 8 mm into the resonator and the resonant frequency

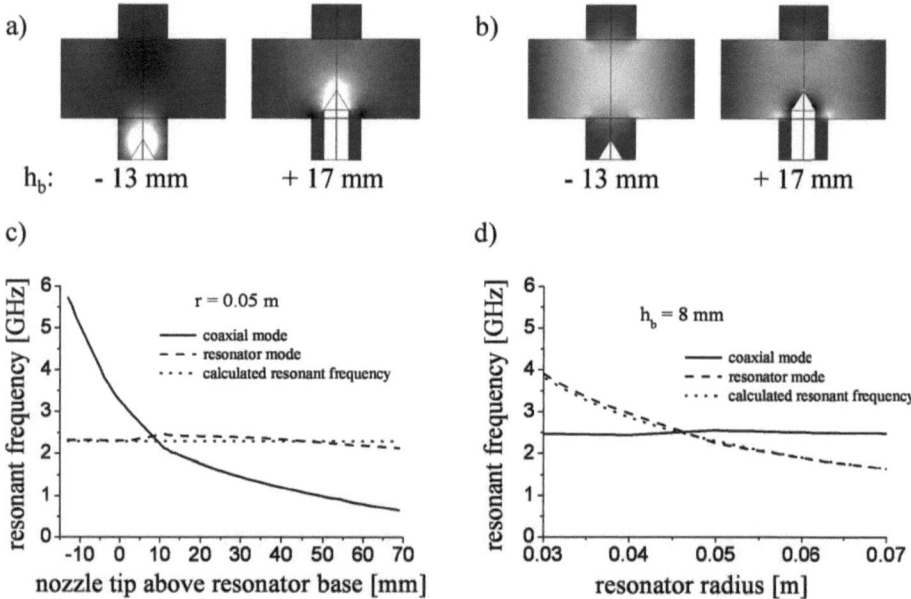

Figure 2.10: Distribution of the z-component of the electric field for the two different nozzle positions and for a) the coaxial and b) the resonator mode. The two diagrams show the resonant frequency in dependence of c) the nozzle position h_b and d) the resonator radius r.

decreases when the nozzle is moved into the resonator. Measurements of the resonant frequency in dependence of the nozzle position were performed to verify this characteristic. The simulated and experimental results are in excellent agreement and are presented in section 2.4.

Since the above presented simulations revealed that the position of the metallic nozzle has a dramatic influence on the resonant frequency as well as on the electric field distribution, simulations of the distribution of all electric field components and of the resonant frequency in dependence of the nozzle position and of the cylindrical resonator radius were performed. These simulations revealed that the APS, which is based on a cylindrical resonator with a metallic nozzle, needed for the gas inlet, actually consists of two resonators: the cylindrical resonator and a second coaxial resonator which is formed below the cylindrical resonator by the metallic nozzle. Moreover, Eigenfrequency analyses showed that two modes exist: the well-known resonator mode and a second mode, which will be called coaxial mode in the following.

Fig. 2.10a) and b) show distributions of the z-components of the electric field of the two different modes for two different nozzle positions of $-13\,\text{mm}$ and $17\,\text{mm}$, which were performed by Eigenmode analyses. Fig. 2.10a) shows the distribution of the z-component of the electric field for the coaxial mode and b) for the resonator mode. The z-component of the resonator mode has its maximum in the centre of the cylindrical resonator for both nozzle positions whereas the coaxial mode reaches its maximum value at the nozzle tip. So the two modes have distinct different distributions of the z-component of the electric field.

The diagrams in Fig. 2.10c) and d) show the dependence of the resonant frequencies of the two modes on the nozzle position and on the resonator radius, respectively. The solid curves show the dependence of the resonant frequency of the coaxial mode. The frequency dependence of the resonator mode is given by the dashed curves, which show the simulated results, and the dotted curves, which display the analytically calculated resonant frequencies by using equation 2.11. It can be seen, that the resonant frequency of the coaxial mode is only dependent on the nozzle position and decreases when the nozzle is moved into the resonator (solid curve in Fig. 2.10c)). However, the resonant frequency of the coaxial mode is not affected by the variation of the resonator radius (solid curve in Fig. 2.10d)). The resonant frequency of the resonator mode is on the other hand independent of the nozzle position (dashed and dotted curve in Fig. 2.10c)) but shows a strong dependence on the resonator radius, which is demonstrated by the simulations as well as by the analytical calculations (dashed and dotted curve in Fig. 2.10d)). This can already be seen, when regarding equation 2.11:

$$\nu_{010} = \frac{\omega_{010}}{2\pi} = \frac{x_{01}}{r\sqrt{\epsilon\mu}},$$

which shows that the resonant frequency is linear in $\frac{1}{r}$ for E_{0n0}-mode resonators.

These simulations showed that the APS is based on two resonators, the cylindrical resonator and the coaxial one, which is formed by the metallic nozzle. This leads to two different modes: the resonator mode and the coaxial mode. The resonator mode has a high electric field in its centre and the resonant frequency is independent of the nozzle position but dependent on the resonator radius. In contrast, the coaxial mode has a maximum electric field at the nozzle tip and its resonant frequency is independent of the resonator radius but dependent on the nozzle position. Thus the resonant frequency of the two modes can be varied on the one hand by the resonator radius for the resonator mode and on the other hand by the nozzle position for the coaxial mode. When the resonator radius and the nozzle position are chosen in such a way, that both modes have the same resonant frequency, the electric field in the APS can be increased even more drastically. This is shown by simulations of an APS, which is driven by microwaves of 915 MHz and was realised for this APS, which was newly designed and manufactured [41].

Further simulations of the electric field distribution as well as Eigenfrequency analyses of this 915 MHz APS where performed concerning the variation of other parameters like the height h_{coax} or the diameter d_{coax} of the coaxial part which revealed their influence on the resonant frequency [41]. These simulations disclosed, that a larger coaxial height leads to a higher electric field and thus to a configuration with a higher quality [41]. This can again be explained by the fact, that the size of the whole APS configuration is large compared to the dimensions of the waveguide.

To summarise, the simulations of a cylindrical resonator with the same dimensions as the APS showed, that such a resonator only has a low quality and according to this a low electric field in its centre, so that a self ignition of the plasma is almost impossible. Therefore, simulations of higher mode resonators such as E_{020}- and E_{030}-mode resonators were performed and showed that these configurations have a higher quality and thereby a higher electric field in their centre since the dimension of the resonator is large compared to the size of the waveguide.

Another possibility to enhance the quality, is to use special coupling elements. Thereby the coupling hole between the waveguide and the resonator is minimised. The simulations of the three coupling elements, a slit, a $\lambda/4$-part, and a taper, displayed that the highest electric field can be achieved when the $\lambda/4$-part is used, whereas the taper has the best microwave transmission properties. A further improvement of the quality can be obtained when a higher mode resonator is combined with a coupling element. To examine if these configurations are suitable for an atmospheric pressure microwave plasma source, which provides self ignition of plasma and stable plasma operation and to verify the simulation results a E_{030}-mode resonator, a slit, a $\lambda/4$-part and a taper were designed, manufactured, and measured by using a network analyser. The experimental results are presented in section 2.4 and compared to the simulations.

Furthermore, the simulations of the real APS with inserted quartz tube revealed that, the inserted quartz tube virtually enlarges the resonator which leads to a shift of the resonant frequency to lower resonant frequencies. The simulations of the APS with the metallic nozzle, which is needed for the gas inlet, showed that the electric field distribution as well as the resonant frequency are affected radically by the metallic nozzle. The resonant frequency decreases when the nozzle is moved into the resonator. On the other hand this means that the nozzle can be used to adjust the resonant frequency and thus functions as a tuning element, which can be used to develop a self igniting atmospheric pressure microwave plasma source, as will be presented in chapter 3. Further simulations and Eigenmode analyses revealed that the nozzle forms a coaxial structure below the cylindrical resonator, so that the APS is actually based on two resonators: the cylindrical resonator and a coaxial one. Eigenfrequency analyses revealed that there are two possible modes: The resonator mode which has a high electric field in the resonator centre and whose resonant frequency is independent of the nozzle position but depends on the resonator radius and on the other hand the coaxial mode with its highest electric field at the nozzle tip. The resonant frequency of the coaxial mode is independent of the resonator radius but it shows a strong dependency on the nozzle position.

Thus the simulations delivered detailed information about the electric field distribution of different kinds of configurations and revealed how higher quality configurations can be attained and fundamentals of the APS, like that it consists of two resonators, what further aided the development of a self igniting APS, which will be presented in chapter 3. However, before that measurements of the microwave properties of the configurations discussed in this section 2.3.2 are presented in the following section 2.4.

2.4 Measurements of the Microwave Characteristics and Comparison with the Simulations

The simulations with COMSOL Multiphysics^{TM} in section 2.3 showed, that configurations with a high electric field in the resonator centre can be achieved when higher mode resonators, for example E_{020}- or E_{030}-mode resonators, and/or special coupling elements, like a taper, a $\lambda/4$-part or a slit, are used which increase the quality of the configuration. To verify these results a E_{030}-mode resonator, a taper, a $\lambda/4$-part and slit were designed and manufactured and are presented in section 2.4.1. The quality of configurations, consisting of the E_{030}-mode resonator or the APS combined with the coupling elements, were measured with the experimental setup presented in section 2.4.2 by using a network analyser. These results are presented in section 2.4.3. Furthermore, Eigenfrequency analyses of the real atmospheric pressure microwave plasma source have been performed and revealed that the quartz tube, which confines the plasma, causes a shift of the resonant frequency to lower frequencies. Simulations of the electric field distributions as well as Eigenfrequency analyses of the APS with the metallic nozzle showed, that the resonant frequency depends on the nozzle position and decreases when the nozzle is moved into the cylindrical resonator. To verify these results from the simulations, measurements of the resonant frequency of the ASP with and without an inserted quartz tube as well as in dependence of the nozzle position were performed and are presented in section 2.4.3.

2.4.1 Design and Construction of the E_{030}-Mode Resonator and the Coupling Elements

The simulation of the electric field distribution revealed, that a high electric field in the centre of the resonator, which is needed for a self ignition of the plasma, can be achieved when either higher mode resonators or special coupling elements are used. The highest electric field can be attained when a higher mode resonator is combined with one of these coupling elements. The simulations showed that the highest electric field can be obtained when the resonator dimension is large compared to the waveguide, which lead to an E_{030}-mode resonator. This resonator is, with a radius of $0.1684\,\text{m}$, comparably large to the waveguide ($a = 0.0864\,\text{m}$ and $b = 0.0432\,\text{m}$) but not too large to be impractical to handle.

To verify the results obtained by the simulations, concerning the quality and the resonant frequency, and to test if the quality, and according to that the electric field in the centre of an E_{030}-mode resonator, is high enough to ignite plasma without any additional igniters, an E_{030}-mode resonator was designed and manufactured.

To confine the plasma again a quartz tube was chosen but with a larger diameter of $0.04\,\text{m}$ (diameter of the quartz tube of the E_{010}-APS: $0.03\,\text{m}$), so that the quartz tube is located in a place, where the electric field is low and according to that only a small shift of the resonant frequency is caused. Additionally, with a larger quartz tube higher gas flows can be treated. With a larger quartz tube even higher gas flows could be treated and this tube could be placed in the second

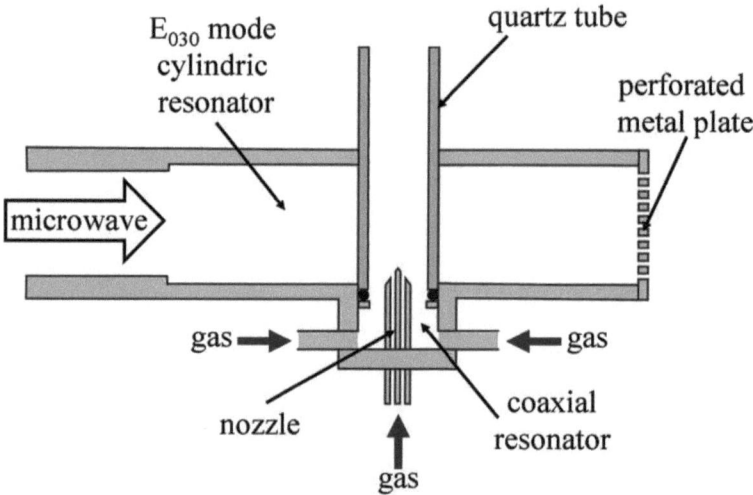

Figure 2.11: Schematic view of the E_{030}-APS with a radius of $r = 0.1634\,\text{m}$ and a height of $h = 0.06\,\text{m}$. The microwave can be coupled in from the left side via a waveguide or a coupling element. The top and bottom of the E_{030}-APS are made of an aluminium plate and the wall is made of a perforated metal plate. Fig. 3.4 in section 3.4 shows a photo of the E_{030}-APS.

minimum of the electric field, so that again the shift of the resonant frequency is small. However, any hole in the resonator, here for the quartz tube, must be smaller than half the wavelength of the microwave power since otherwise microwave would be coupled out through this hole. The gas inlet was again realised with a metallic nozzle, so that this atmospheric pressure microwave plasma source, also consists of two resonators and therefore has a resonant frequency tuning element, the metallic nozzle. Further on, this APS, which is based on a E_{030}-mode resonator, will be named E_{030}-APS while the E_{010}-mode-based APS will be named E_{010}-APS.

The radius of the E_{030}-APS was calculated iteratively by simulations. Therefore, Eigenfrequency analyses with COMSOL Multiphysics^{TM} of the E_{030}-APS with the inserted quartz tube and one fixed nozzle configuration were performed iteratively for different radii until the resonant frequency of the E_{030}-APS coincided with the magnetron frequency of $\nu = 2.45\,\text{GHz}$. This led to a radius of $r = 0.1634\,\text{m}$ for the E_{030}-APS. Thus the E_{030}-APS was manufactured with this radius and a height of $0.06\,\text{m}$. Fig. 2.11 shows a schematic view of the E_{030}-APS. The top and the bottom of the E_{030}-APS resonator are made of aluminium plates whereas the wall is made of a perforated metal plate, so that the plasma inside the resonator is accessible for diagnostics. The microwave can be coupled in from the left side via a waveguide or via a coupling element. In addition to the central gas inlet through the metallic nozzle, a second tangential gas inlet was integrated, which enhances a stable plasma operation. To verify the simulation results, the microwave properties, like the quality and the resonant frequency, of the E_{030}-APS were characterised by using a network analyser and will be presented in section 2.4.3. Plasma operation was also tested and will be

2.4. COMPARISON OF MEASUREMENTS AND SIMULATION

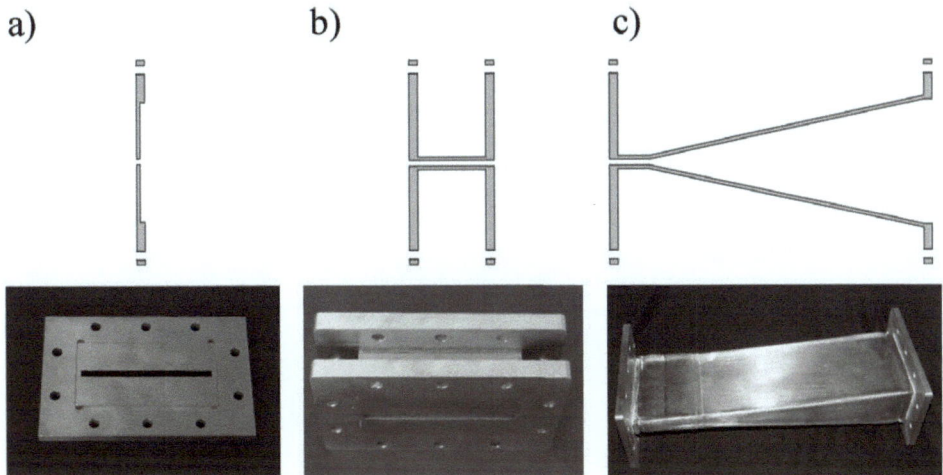

Figure 2.12: Schematic view and photos of the three coupling elements: a) a slit with a thickness of $d = 1\,\text{mm}$ and a height of $h = 5\,\text{mm}$, b) a $\lambda/4$-part with a length of $l = \lambda_h/4 = 43\,\text{mm}$ and a height of $h = 5\,\text{mm}$, and c) a taper with a tapered length of $l = 1.5\lambda_h = 0.258\,\text{m}$ and a tapered height of $5\,\text{mm}$.

presented in chapter 3.

The simulations presented in the previous section 2.3 showed that a higher quality can not only be achieved by using a higher mode resonator but also when special coupling elements are used. Three coupling elements were analysed in section 2.3: a slit, a $\lambda/4$-part, and a taper. Fig. 2.12 shows schematic views and photos of the manufactured three coupling elements. The slit has a thickness of $d = 1\,\text{mm}$ and a height of $h = 5\,\text{mm}$. The $\lambda/4$-part has a length of $l = 43\,\text{mm}$ and the same height as the slit: $h = 5\,\text{mm}$. The taper has the same tapered height and a tapered length of $1.5\lambda_h = 0.258\,\text{m}$. Their microwave properties, here the improvement of the quality when they are combined with an APS, were analysed with a network analyser and the results will be presented in section 2.4.3.

2.4.2 Experimental Setup

The simulations of the electric field distribution revealed that the quality, and according to that the electric field, can be enhanced when higher mode resonators or special coupling elements are used. The Eigenmode analyses showed, that the resonant frequency of the E_{010}-APS is dependent on the position of the metallic nozzle and decreases when the nozzle is moved into the resonator. So to verify the results obtained by the simulations, measurements of the resonant frequency as well as measurements of the quality of the sole E_{010}- and E_{030}-APS and combined with the coupling elements were performed.

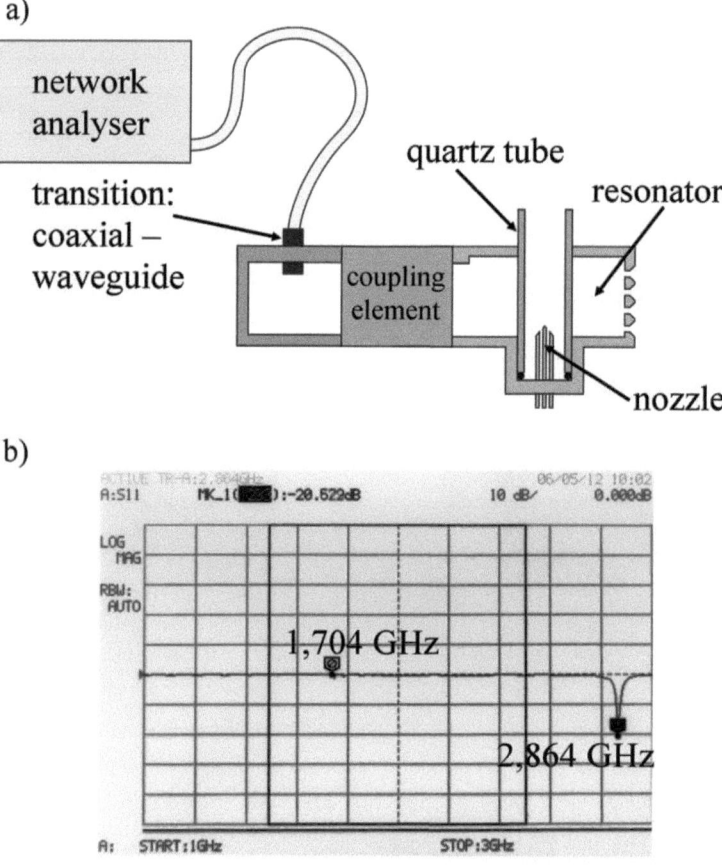

Figure 2.13: a) A schematic view of the experimental setup for the measurement of the microwave properties using a network analyser and b) a typical measurement. The marker on the left denotes the cutoff frequency of the rectangular waveguide while the right marker denotes the dip at the resonant frequency.

The microwave properties of such configurations can be characterised by the scattering parameters, the so called S-parameters [55]. S-parameters are commonly used to describe the behaviour of linear networks in dependence of the frequency. The number of required S-parameters is dependent on the number of gates and is given by the square of the gates. A configuration which consists of a resonator coupled to a waveguide, like the APS, is a oneport, has only one gate, namely the front of the waveguide where the microwave is coupled in, and therefore is characterised by only one S-parameter.

A network analyser was used to measure the S-parameters of the E_{010}- and the E_{030}-APS and of the coupling elements. The utilised MS4662A network analyser from Anritsu can be applied to characterise two-port networks, which have two gates. A twoport can be characterised by four S-parameters: S_{11}-, S_{21}-, S_{21}-, and S_{22}-parameter. The S_{11}- and S_{22}-parameter characterise the reflection properties while the S_{12}- and S_{21}-parameter characterise the transmission properties. Since a one-port, the E_{010}- and E_{030}-APS (combined with the coupling elements), is measured here, only the S_{11}-parameter was explored. The S_{11}-parameter expresses, as already mentioned, the reflection coefficient. The reflection coefficient is dependent on the frequency of the incident microwave. When the resonator is stimulated at its resonant frequency only small amounts of the microwave power are reflected back into the waveguide. This means, that the reflection coefficient and according to that the S_{11}-parameter, has a minimum, a dip, at the resonant frequency. Thus, when the S_{11}-parameter of an APS is measured in dependence of the frequency, the resonant frequency can be determined from the position of the minimum. Furthermore, the width of that dip provides information about the quality of the measured configuration. A wide dip denotes a low quality while a narrow dip indicates a high quality of the configuration.

Fig. 2.13a) shows the schematic view of the experimental setup for the measurement of the microwave properties. A measuring signal of a few mW is produced by the network analyser and guided to the coaxial-to-waveguide transition via a coaxial cable where the microwave is coupled into the rectangular waveguide and guided to the resonator. When configurations with coupling elements were measured, the coupling elements were placed between the coaxial-to-waveguide transition and the resonators. The reflected signal is detected by the network analyser, the S_{11}-parameter is calculated and displayed. The frequency of the measuring signal was varied from 1 GHz to 3 GHz. Fig. 2.13b) shows a typical measurement of the S_{11}-parameter. The marker on the left denotes the cutoff frequency of the rectangular waveguide $\nu_{cutoff} = 1.704$ GHz while the right marker denotes the dip at the resonant frequency, here $\nu_{res} = 2.864$ GHz.
So the microwave properties of the E_{010}- and E_{030}-APS as well as of the coupling elements, the resonant frequency and the quality, can be measured with this experimental setup using a network analyser. The resonant frequency can be determined from the position of the resonant dip of the S_{11}-parameter while the quality is given by the width of the resonant dip. The following section 2.4.3 presents measurements of the resonant frequency and of the quality of the different configuration and compares them to the simulation results.

2.4.3 Experimental Results

The simulations in section 2.3.2 revealed that configurations with a higher quality, and therewith associated a high electric field in the resonator centre, can be achieved when higher mode resonators or special coupling elements are used. Therefore, a E_{030}-mode resonator and the three coupling elements were designed and manufactured as described in section 2.4.1. The quality of different configurations was measured via the experimental setup presented in section 2.4.2. In the experiment the Full Width at Half Maximum FWHM of the resonant dip is assumed as a measure of the quality. In the simulations the quality is displayed by the norm of the electric field in the

Table 2.1: This table summarises the quality of the sole E_{010}- and E_{030}-APS and configurations with the three coupling elements, the slit, the taper, and the $\lambda/4$-part, in combination with the APS. As a measure of the quality of the configurations the norm of the electric field in the resonator centre E_{centre} is assumed for the simulations whereas the Full Width at Half Maximum (FWHM) is assumed for the experiment.

	simulation norm of the el. field in the resonator centre E_{centre} [V/m]	experiment Full Width at Half Maximum FWHM [MHz]
E_{010}	$0.73 \cdot 10^5$ V/m	76.0 MHz
E_{010} with slit	$1.04 \cdot 10^5$ V/m	22.0 MHz
E_{010} with taper	$2.85 \cdot 10^5$ V/m	10.6 MHz
E_{010} with $\lambda/4$-part	$3.56 \cdot 10^5$ V/m	6.40 MHz
E_{030}	$1.39 \cdot 10^5$ V/m	7.96 MHz
E_{030} with slit	$1.77 \cdot 10^5$ V/m	3.36 MHz
E_{030} with taper	$5.80 \cdot 10^5$ V/m	2.92 MHz
E_{030} with $\lambda/4$-part	$18.4 \cdot 10^5$ V/m	2.64 MHz

resonator centre E_{centre}. The quality of the sole E_{010}-APS with a radius of $r = 0.05$ m and the E_{030}-APS, described in section 2.4.1, as well as a combination of the APS with the coupling elements (slit, taper, and $\lambda/4$-part), were measured. Table 2.1 summarises these measurements and compares them to the simulation results. The simulations revealed an E_{centre} of the sole E_{010}-APS of $0.73 \cdot 10^5$ V/m whereas the measurements revealed a FWHM of 76.0 MHz. When the E_{010}-APS is combined with a slit, the E_{centre} increases to $1.04 \cdot 10^5$ V/m and the FWHM decreases to 22.0 MHz. So the simulation as well as the measurements show that the quality of the E_{010}-APS combined with a slit has increased compared to the sole E_{010}-APS. When a taper instead of the slit is used the quality can be improved as $E_{centre} = 2.85 \cdot 10^5$ V/m and FWHM = 10.6 MHz show. A further improvement can be achieved when a $\lambda/4$-part is used instead which leads to an electric field of $3.56 \cdot 10^5$ V/m and to a Full Width at Half Maximum of 6.40 MHz. This displays that for the E_{010}-APS the simulations and the measurements show the same behaviour. The simulation and the measurement revealed an E_{centre} of $1.39 \cdot 10^5$ V/m and a FWHM of 7.96 MHz, respectively, for the sole E_{030}-APS which means that the quality has increased compared to the E_{010}-APS. Again configurations of the E_{030}-APS combined with the slit, the taper, or the $\lambda/4$-part were simulated and measured. These simulations and measurements showed, that the quality can be improved when a slit is used. A further improvement can be obtained when a taper is used and the best results can be achieved when a $\lambda/4$-part is combined with an E_{030}-APS.

So the measurements show the same behaviour as the simulations and illustrate that the quality can be improved when higher mode resonators and special coupling elements are used.

2.4. COMPARISON OF MEASUREMENTS AND SIMULATION

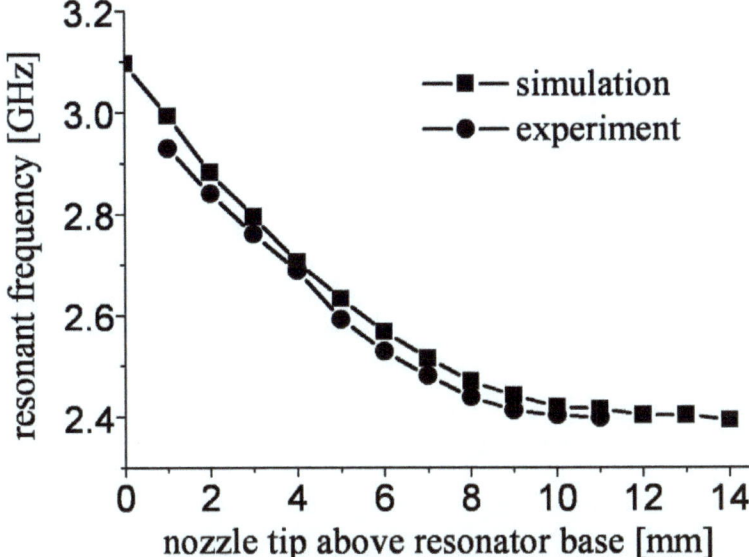

Figure 2.14: Dependency of the resonant frequency on the nozzle position. The square symbols were obtained by the simulations while the round symbols represent the measured values. Both curves show the same behaviour and are in excellent agreement: the resonant frequency decreases when the nozzle is moved into the resonator.

The simulations in section 2.3.2.3 showed that the inserted quartz tube, which is needed to confine the plasma, causes a shift of the resonant frequency to lower frequencies. Therefore, measurements of the resonant frequency with the network analyser of the E_{010}-APS with and without inserted quartz tube were performed. The S_{11}-parameter was measured and the resonant frequency was determined from the position of the resonant dip. For the E_{010}-APS the resonant frequency was determined to 2.736 GHz without quartz tube and 2.584 GHz with the quartz tube for a fixed nozzle position. Thus the simulations and measurements agree that inserting a quartz tube into the resonator causes a shift in the resonant frequency to lower frequencies.

Furthermore, the simulations in section 2.3.2.3 revealed, that the resonant frequency of the E_{010}-APS decreases when the metallic nozzle, which is used for the gas inlet, is moved into the resonator. To measure the resonant frequency in dependence of the nozzle position the E_{010}-APS was equipped with a movable nozzle. Fig. 2.14 shows the resonant frequency in dependence of the nozzle position. The square symbols were obtained by the simulations performed in section 2.3.2.3 while the round symbols show the measured values. The curves show an excellent agreement between the simulation and the measurement. Since the resonant frequency is dependent of the nozzle position the nozzle can be used as a tuning element to adjust the resonant frequency. This leads to an E_{010}-APS configuration, which provides self ignition of the plasma and maintains

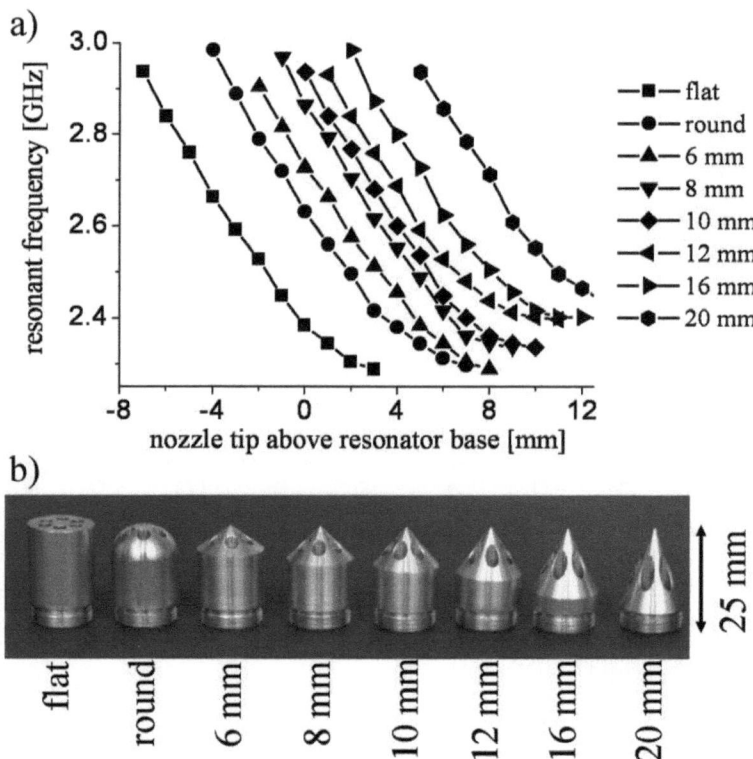

Figure 2.15: a) Dependence of the nozzle position on the resonant frequency for differently shaped nozzle tips. b) Photo of the different nozzle tips.

stable plasma operation as will be presented in chapter 3.

Moreover, differently shaped nozzle tips were manufactured to analyse the influence on the resonant frequency. Fig. 2.15b) shows a photo of the different nozzle tip shapes which include a range of pointed tips shown on the right side of the photo, a round nozzle tip and a completely flat tip seen on the left side of the photo. The pointedness of the tip is characterised by the height of the pointed part of the tip and ranges from 20 mm, a relatively strongly pointed tip, to 6 mm, which is only slightly pointed, as can be seen in the photo. The diagram in Fig. 2.15a) shows the dependence of the resonant frequency on the nozzle position for the different nozzle tips. The dependence is the same for all the nozzle tips: the resonant frequency decreases when the nozzle is moved into the resonator. However, the resonant frequency for a fixed nozzle position shifts for the different nozzle tips. For example, a resonant frequency of 2.94 GHz can be achieved when the nozzle with the flat tip extends 8 mm into the resonator whereas the same resonant frequency is obtained when the nozzle with the round tip extends about 3.5 mm into the resonator. A relatively

2.4. COMPARISON OF MEASUREMENTS AND SIMULATION 55

sharp nozzle tip, e. g. the 20 mm-nozzle has a resonant frequency of 2.712 GHz when the nozzle extends 8 mm into the resonator while a relatively flat tip, like the 6 mm-nozzle, has a distinct lower resonant frequency of 2.288 GHZ for the same nozzle position. This means that a more pointed nozzle tip leads to a higher resonant frequency compared to nozzles with a less pointed tip in the same position.

To summarise, the simulations in section 2.3 revealed, that a sole E_{010} resonator coupled to a waveguide only has a very poor quality and therefore, the electric field in the resonator centre is not higher compared to the one in the waveguide. This led to configurations with higher qualities: higher mode resonators and special coupling elements. To verify the simulation results an E_{030}-mode based APS as well as the three analysed coupling elements, the slit, the taper, and the $\lambda/4$-part, were designed and manufactured as described in section 2.4.1.
The measurements of the quality of the individual configurations showed that the sole E_{010}-APS has the lowest quality. The quality can be improved when a slit is used as coupling element. Further improvements can be achieved when the taper or the $\lambda/4$-part is used. The measurements of the quality of the E_{030}-APS showed that the quality can also be improved when higher mode resonators are used and an even higher improvement can be achieved when the E_{030}-APS is combined with the coupling elements. So the experimental and simulation results agree.
The simulations also showed that the inserted quartz tube, which is needed to confine the plasma, causes a shift of the resonant frequency to lower resonant frequencies. This could be verified by measurements of the resonant frequency of the E_{010}-APS with and without an inserted quartz tube.
Further simulations revealed, that the resonant frequency is dependent on the position of the metallic nozzle, which is necessary for the gas inlet. Therefore, the resonant frequency of the E_{010}-APS was measured in dependence of the nozzle position and showed, that the simulated and measured results are in excellent agreement. Further on, measurements of the resonant frequency in dependence of different nozzle shapes were performed. These revealed that the dependence of the nozzle position on the resonant frequency is the same for all analysed nozzle shapes, in that the resonant frequency decreases when the nozzle is moved into the resonator. However, the shape of the nozzle tip has an influence on the resonant frequency: More pointed nozzle tips lead to a higher resonant frequency than nozzle tips, which are less pointed, for the same nozzle position.
As already mentioned above, the nozzle position depends on the resonant frequency, which can be measured with a network analyser. Regarded from a different perspective, this means that the resonant frequency of the APS can be adjusted by the position of the metallic nozzle. This fact leads to an APS configuration which is able to ignite plasma without any additional igniters and maintains stable and continuous plasma operation. This APS will be presented in the following chapter 3.

Chapter 3

The Self Igniting APS

The atmospheric pressure microwave plasma source APS is based on a cylindrical resonator as described in section 2.1. The plasma is confined in a quartz tube. A metallic nozzle, which is needed for the gas inlet, forms a coaxial structure below the cylindrical resonator. The simulations in chapter 2.3 revealed, that the quartz tube causes a shift of the resonant frequency and that the metallic nozzle affects the resonant frequency of the APS. The measurements in section 2.4 verified that the resonant frequency decreases when the nozzle is moved into the resonator. Since the magnetron frequency is fixed, but varies in dependence of the type and on the output power, the resonant frequency of the APS can be adjusted to the frequency of the microwave generated by the magnetron. To determine the frequency of the produced microwave the magnetron was measured with a spectrum analyser, which is described in the following section 3.1. In combination with a commonly used setup for the operation of microwave-generated plasma sources, which is described in section 3.2, this leads to a self igniting E_{010}-APS, which is presented in section 3.3. Section 3.4 deals with a self igniting E_{030}-APS.

3.1 Measurements of the Magnetron

To acquire knowledge about the frequency dependency of the utilised magnetron the frequency of the radiated microwave was measured in dependence of the output power with a spectrum analyser. The output power of the utilised magnetron ranges from 0.3 kW to 3 kW. Fig. 3.1 shows the frequency of the radiated microwave in dependence of the output power. The diagram illustrates, that the frequency of the produced microwave increases from 2.4515 GHz to 2.4563 GHz when the output power of the magnetron is increased from 0.3 kW to 2.7 kW. When the output power is further increased to 3 kW the frequency of the radiated microwave decreases somewhat to 2.4560 GHz. Since now the frequency dependency of the magnetron is known the resonant frequency of the APS can be adjusted to the frequency of the microwave generated by the magnetron.

3.2. COMMONLY USED EXPERIMENTAL SETUP

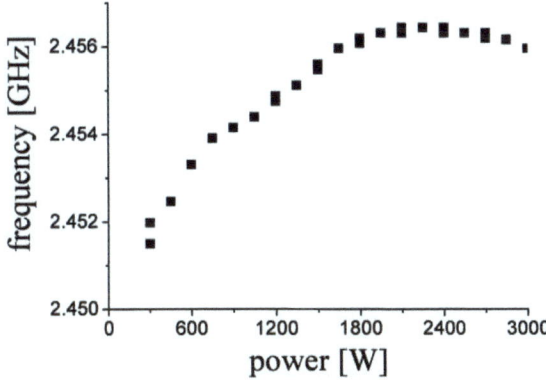

Figure 3.1: Dependency of the microwave frequency on the output power of the magnetron.

3.2 Commonly used Experimental Setup for the Operation of Microwave-generated Plasma Sources

Commonly an experimental setup as shown in Fig. 3.2 is used for the operation of microwave-generated plasma sources. The plasma source is supplied with microwave power of 3 kW, which is produced by the magnetron. The magnetron is shown in Fig. 3.2 on the left side. The circulator located to the right of the magnetron protects the magnetron from microwave power reflected by

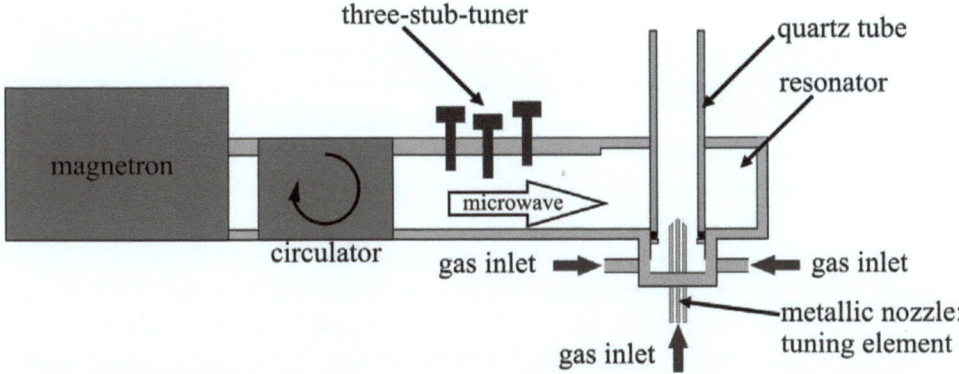

Figure 3.2: Commonly used experimental setup for the operation of microwave-generated plasma sources. The microwave is produced by the magnetron, which is shown on the left side. Then the microwave is guided through a circulator with a waterload and a three-stub-tuner to the plasma source via waveguides, shown on the right side. The circulator is used to protect the magnetron from reflected microwave power whereas the three-stub-tuner is used for impedance matching.

the plasma source. The reflected microwave is redirected into the waterload where it is dumped. This is necessary because microwave reentering the magnetron would destroy it.

A well known impedance matching element is a so called three-stub-tuner. A three-stub-tuner consists of a part of a waveguide with three stubs at a distance of about $3/8\lambda_h$, which can be moved into the waveguide. This device can be used to match the impedance from the waveguide to the plasma source. It can be achieved by placing the three stubs in the waveguide in such a way, that the reflected microwave power from the plasma source is reflected back to the plasma at the stubs, so that the forward power to the plasma source in maximised. Behind the three-stub-tuner the plasma source is mounted, here the APS.

3.3 The Self Igniting E_{010}-APS

The simulations as well as the measurements of the microwave properties revealed that the metallic nozzle affects the resonant frequency of the APS. The resonant frequency decreases when the nozzle is moved into the resonator. Thus the resonant frequency of the APS can be tuned by the position of the metallic nozzle to match the frequency of the microwave, which is generated by the magnetron and already known by measurement. This can be achieved with the same experimental setup, which was used to measure the microwave properties and is described in section 2.4.2.
The E_{010}-APS with an inserted quartz tube, coupled to a three-stub-tuner, is connected to the network analyser and the resonant frequency is determined from the position of the resonant dip in the S_{11}-parameter. The nozzle is moved up and down iteratively until the resonant frequency of the APS has the same value as the microwave generated by the magnetron. Then the impedance is matched with the three-stub-tuner by iteratively adjusting the three stubs, so that the resonant dip gets as deep as possible, which implies that the forward power is maximised. After this procedure the whole E_{010}-APS with the positioned three stub tuner is mounted behind the magnetron and the circulator. Then the magnetron is turned on and a microwave power of about only 0.3 kW..0.5 kW is sufficient to ignite a plasma in the E_{010}-APS without any additional igniters. After the plasma is ignited the three-stub-tuner must be readjusted since the impedance of the APS has changed due to the plasma burning in it. Then the microwave can be supplied continuously to the plasma and is almost completely absorbed by the plasma.
The fact that an E_{010}-APS, whose resonant frequency is adjusted to the frequency of the microwave, in combination with a three-stub-tuner provides a sufficiently high electric field to ignite plasma without any additional igniters can be explained by the fact, that the three-stub-tuner acts as an excellent coupling element. Since the three-stub-tuner reflects the microwave, which is reflected from the APS, back into the resonator, the three-stub-tuner enhances the quality, so that a high electric field can be established in the resonator centre, which is sufficiently high to ignite plasma without any additional igniters. Thus, the three-stub-tuner acts as an excellent coupling element. Therefore, it would be very interesting to simulate the electric field distribution of such a configuration consisting of an APS combined with a three-stub-tuner and to compare these simula-

3.3. THE SELF IGNITING E_{010}-APS

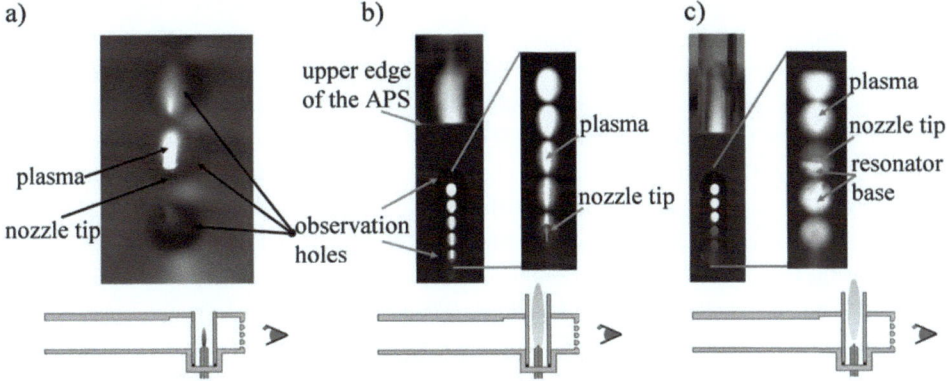

Figure 3.3: Different kinds of plasma modes in the E_{010}-APS. In the upper part of the figure photos of the different plasma modes are shown and below schematic views illustrate how the different plasma modes are located in the APS. a) A small plasma flame at the nozzle tip, which reflects the electric field distribution of the coaxial mode, b) an extended plasma in the centre of the resonator with contact with the nozzle tip, and c) a free-standing, extended plasma in the resonator centre, without contact with the nozzle, which both reflect the electric field distribution of the resonator mode.

tions to the simulations presented in earlier chapters. However, the iterative process of finding the optimal stub positions would have to be performed manually which was deemed to time-consuming.

In summary the development-history of a self-igniting E_{010}-APS was as follows. First the simulations of the realistic E_{010}-APS revealed that the resonant frequency is dependent on the nozzle position and decreases when the nozzle is moved into the resonator, which was verified by measurements. On the other hand this led to the fact that the nozzle provides a tuning element, which can be used to adjust the resonant frequency of the APS to the frequency of the utilised microwave. The use of a three-stub-tuner can improve the quality of the E_{010}-APS to such an extent that the electric field in its centre, which is established by microwaves of a power of about only 0.3 kW..0.5 kW, is sufficiently high to ignite a plasma in air at atmospheric pressure without any additional igniters.

The Different Plasma Modes

The plasma which can be ignited in the APS can assume different shapes. Fig. 3.3 shows photos and schematic views of the different plasma shapes. The plasma is observed via five small holes in the front of the APS. The photo in Fig. 3.3a) shows a small plasma at the nozzle tip. The plasma forms a converging filamentary cone, which is delimited to a small region above the nozzle tip. This plasma mode reflects the electric field distribution of the coaxial mode shown in section 2.3.2.3, which has a high electric field at the nozzle tip. This coaxial plasma mode can be observed at very high gas flows when the supplied microwave power is comparably low.

When gas flows, which are not too high, are used and enough microwave power is supplied, the plasma jumps into another mode and an extended plasma can be observed. The plasma nearly fills the complete quartz tube and protrudes above the cylindrical resonator. The plasma still has contact with the nozzle tip and reflects the electric field distribution of the resonator mode, which was shown in section 2.3.2.3. Fig 3.3b) shows a photo and a schematic view of this plasma mode. The photo on the left shows the whole APS with the plasma, which protrudes above the upper edge of the APS. The right photo shows a close look at the plasma through the five holes and it can be seen that the plasma is in contact with the nozzle tip.

An extended plasma, which has no contact with the nozzle and therefore is free-standing, can also be attained if the gas flow as well as the microwave power are not too high. This is illustrated by the photo and the schematic view in Fig. 3.3c). The left photo again shows the whole APS. On the right photo the plasma is observed through the five holes and it can be seen that the plasma has no contact with the nozzle. Since in this mode the plasma has no contact with the nozzle it has the advantage that the nozzle cannot be damaged by the plasma. The extended plasmas also have the advantage that larger amounts of waste gases can be treated, since the plasma almost fills the complete quartz tube and is also prolonged in the axial direction. A closer look at the plasma shape in dependence of the gas flow and of the microwave power is presented in section 4.2.3 and their effect on the decomposition of waste gases is presented in section 4.3.

3.4 The Self Igniting E_{030}-APS

It was presented, in the previous section 3.3, that already the E_{010}-APS in combination with a three stub tuner provides a self igniting microwave plasma source at atmospheric pressure. This can be achieved by using the metallic nozzle as a tuning element to adjust the resonant frequency to the frequency of the microwave generated by the magnetron. The simulations in section 2.3 showed that the sole E_{010}-APS only has a low quality, which results in a small electric field in its centre. However, the simulations of higher mode based APS showed that these configurations provide a higher quality and therefore a higher electric field in their centre. Therefore, a E_{030}-APS was designed and manufactured as presented in section 2.4. Since this E_{030}-APS has an inherent higher quality compared to the E_{010}-APS, the ignition of a plasma in this E_{030}-APS should be even easier. Therefore, the capability of the E_{030}-APS as a self igniting microwave plasma source at atmospheric pressure and for stable, continuous, and efficient plasma operation was tested.

The frequency dependence of the magnetron has already been measured and presented in section 3.1 because detailed information about the frequency of the microwave was also essential for the plasma ignition and operation of the E_{010}-APS. Since the E_{030}-APS as well as the E_{010}-APS is based on a cylindrical resonator and on a second coaxial one, which is formed by the metallic nozzle, needed for the gas inlet, the resonant frequency of the E_{030}-APS can again be adjusted by this nozzle. The same procedure to adjust the resonant frequency of the E_{030}-APS as was used for the E_{010}-APS, which is presented in the previous section 3.3, is used. A three-stub-tuner, which is used to match the impedance between the waveguide and the APS, is used once again. After the

3.4. THE SELF IGNITING E_{030}-APS

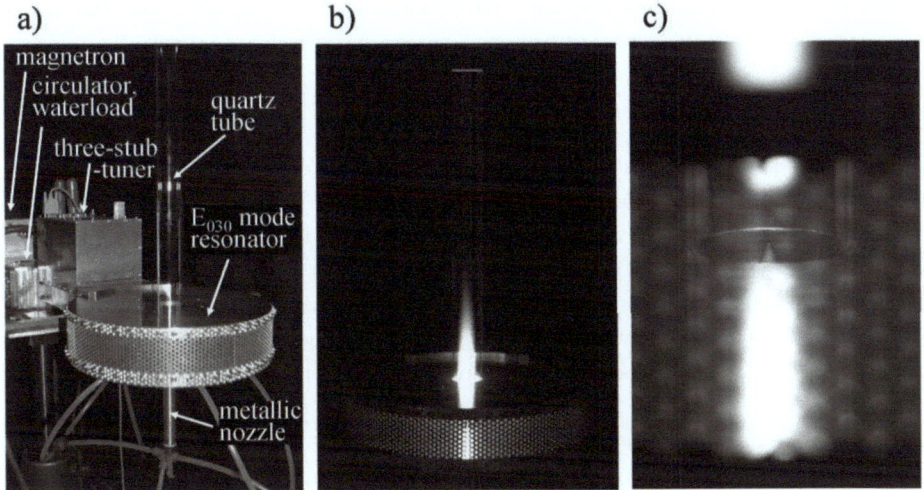

Figure 3.4: a) A photo of the E_{030}-APS with an automatic three-stub-tuner and the circulator with the waterload behind the three-stub-tuner. In the background the magnetron can be seen. The quartz tube protrudes out of the cylindrical resonator. Below the APS a part of the movable nozzle can bee seen as well as the gas supply. The walls of the E_{030}-APS are made of a perforated metal plate, to have good diagnostic access to the plasma inside the resonator, as described in section 2.4. b) A photo of the E_{030}-APS with an extended plasma burning in its centre. c) A close look through the perforated metal plate in the resonator: It can be seen that again a free-standing plasma which has no contact with the nozzle can be generated.

resonant frequency of the E_{030}-APS is adjusted by the nozzle to the frequency of the microwave, the whole configuration is mounted behind the magnetron and the circulator. When the magnetron is turned on, a plasma can be ignited in the E_{030}-APS without any additional igniters. In contrast to the E_{010}-APS, the microwave power which is needed to ignite a plasma, differs most extensively in dependence of very slight deviations of the exact nozzle position.

Fig. 3.4a) shows the E_{030}-APS with an automatic three-stub-tuner and the circulator with the waterload behind the three-stub-tuner. The magnetron can be seen in the background. The photo in Fig. 3.4b) shows an extended plasma burning in the centre of the E_{030}-APS. As with the E_{010}-APS, the three different plasma modes, a small plasma at the nozzle tip, an extended plasma which still has contact with the plasma, and a free-standing, extended, and contactless plasma, were also observed in the E_{030}-APS. A close look through the perforated metal plate in the resonator is shown in Fig. 3.4c). It can be seen that a free-standing plasma with no contact with the nozzle is again possible with this E_{030}-APS plasma source.

The above mentioned fact that the microwave power, which is needed to ignite the plasma, differs that much in dependence of very slight deviations of the exact nozzle position, is caused by the higher quality of this E_{030}-APS configuration compared to an E_{010}-APS. Since the E_{030}-APS has a higher quality Q the resonant frequency is sharper according to equation 2.21:

$$Q = \frac{\omega}{\Delta \omega}.$$

Therefore, the resonant frequency of the E_{030}-APS must coincide precisely with the frequency of the microwave. Since the frequency of the microwave differs in dependence of the output power, as presented in section 3.1, the two frequencies coincide only at a certain output power. This leads to the fact that the power needed for the plasma ignition differs most extensively in dependence of slight deviations of the exact nozzle position. Thus the higher quality of the E_{030}-APS, in comparison to the E_{010}-APS, which was supposed to be an advantage, since it provides a higher electric field in the resonator centre and therefore should facilitate plasma ignition, has turned into a disadvantage.

Another disadvantage is discovered when the three-stub-tuner is readjusted, after the plasma is ignited, to match the changed impedance of the E_{030}-APS with the burning plasma to the waveguide. When the E_{010}-APS was used in the previous section 3.3 nearly the complete supplied microwave power was absorbed by the plasma. Here, when the E_{030}-APS is used, only about half of the supplied microwave power can be absorbed by the plasma. This, as well as the fact, that the resonant frequency of the E_{030}-APS and the frequency of the microwave must precisely coincident, is caused by the high quality of the E_{030}-APS which in turn leads to a sharp resonant frequency. Since the plasma causes a shift of the resonant frequency of the E_{030}-APS due to its dielectric properties, the resonant frequency and the frequency of the microwave no longer match. In case of the E_{010}-APS the resonant frequency was wide enough, due to the lower quality of the E_{010}-APS, so that the resonant frequency still overlaps with the frequency of the microwave. When in contrast the E_{030}-APS is used with an obvious higher quality and therefore a sharper resonant frequency, this resonant frequency overlaps scarcely with the frequency of the microwave. Thus the microwave can no longer be coupled into the E_{030}-APS and is reflected. This can only be partly compensated by the three-stub-tuner. Since the enhanced quality of the E_{030}-APS causes problems at the ignition of the plasma as well as during plasma operation, APS configurations which consist of an APS in combination with a coupling element were not tested, since these configurations provide even higher qualities and therefore would intensify the described problems even more.

These problems, which are found at configurations with a high quality, were also observed by Pott [19]. He used a resonator configuration, which is described in section 2.1, with an enormous quality and had the same problems. Since he used a system which is based on a sole E_{010}-mode resonator, he had no tuning element to adjust the resonant frequency to the frequency of the microwave produced by the magnetron. Therefore, he had to force the magnetron frequency to the resonant frequency by complex injection locking of the magnetron. Aside from that, he also had to face the problem, that the resonant frequency of the resonator is detuned by the burning plasma and then none of the supplied microwave power can be coupled into the resonator. This

3.4. THE SELF IGNITING E_{030}-APS 63

fact induced him to operate this atmospheric pressure microwave plasma source only in pulsed mode. In pulsed mode the microwave can first couple into the resonator, produce a high electric field which is sufficient to ignite a plasma at which point microwave power is no longer coupled into the resonator and the microwave power supply is stopped. The plasma is extinguished, the magnetron is switched on again, and the whole procedure is repeated.

To summarise, the simulations of the realistic E_{010}-APS in section 2.3.2.3 revealed that the resonant frequency is dependent on the position of the metallic nozzle, which is needed for the gas inlet, and decreases when it is moved into the resonator. This could be verified by measurements of the resonant frequency in dependence of the nozzle position. Thus the nozzle position can be used as a tuning element to adjust the resonant frequency of the E_{010}-APS to the magnetron frequency, which is exactly known by measurements with a spectrum analyser. In combination with a three-stub-tuner this results in a microwave plasma source which provides plasma ignition without any additional igniters in air at atmospheric pressure. A three-stub-tuner is commonly used to match impedances. Here it furthermore acts as a coupling element between the waveguide and the E_{010}-APS. Then only a few hundred watts are needed to ignite a plasma in air at atmospheric pressure in the E_{010}-APS without any additional igniters. Three different plasma modes can be observed: a small plasma at the nozzle tip, an extended plasma, which is still in contact with the nozzle, and a free-standing extended plasma above the nozzle.

Since the simulations in section 2.3.2.1 showed that APS, which are based on higher mode resonators, have a higher quality and therefore provide a higher electric field in their centre, they should facilitate the ignition of the plasma. Therefore, an E_{030}-APS was designed and manufactured as described in section 2.4.1. However, the high quality of the E_{030}-APS turned out to be of no advantage. On contrary, the high quality led to disadvantages of the E_{030}-APS. Since the E_{030}-APS has a high quality the resonant frequency is relatively sharp, so that the frequency of the microwave and the resonant frequency must coincide precisely. Furthermore, when the plasma is ignited, the resonant frequency of the E_{030}-APS is shifted by the plasma and the resonant frequency overlaps only slightly with the frequency of the microwave and the microwave can hardly couple into the resonator. Thus only a fraction of the supplied microwave power can be absorbed and the plasma operation becomes inefficient.

So the E_{010}-APS provides an optimal configuration since it offers a sufficiently high quality, which provides a sufficiently high electric field for the ignition of a plasma in air at atmospheric pressure without any additional igniters, but whose quality is not too high and therefore the resonant frequency is wide enough, that, when the plasma is burning, the supplied microwave power still couples into the E_{010}-APS and can be absorbed by the plasma. It offers the possibility to operate in the described three different plasma modes. So the E_{010}-APS combines both of these desirable properties, in that it provides self ignition of the plasma and stable, efficient and continuous plasma operation. Since the E_{010}-APS provides the optimal configuration as a microwave plasma source at atmospheric pressure, the plasma of this E_{010}-APS is characterised and its capability for the decomposition of waste gases is analysed in section 4.2 and 4.3, respectively.

Chapter 4

Characterisation of the Self Igniting APS

In the previous chapters the development and optimisation of the atmospheric microwave plasma source APS, which provides plasma ignition without any additional igniters as well as stable plasma operation, was presented. Now the suitability of the APS for the decomposition of waste gases is analysed in this chapter. Since for a comprehensive understanding of the abatement processes detailed knowledge about the temperatures and densities of the plasma particles is necessary, the APS plasma is characterised first. Thereafter, studies concerning the decomposition of waste gases follow.
So in this chapter at first the experimental setups are presented in section 4.1 and afterwards the characterisation of the APS plasma by means of optical emission spectroscopy as well as analyses concerning the abatement of waste gases are presented in section 4.2 and section 4.3, respectively.

4.1 The Experimental Setup

The characterisation of the APS plasma is performed by using optical emission spectroscopy whose advantages will be presented later on. For the decomposition of waste gas studies the raw and clean gases are characterised by a variety of different measurement devices. Fig. 4.1 shows a schematic view of the complete experimental setup, which is used for the characterisation of the plasma as well as for the analyses of the decomposition of waste gases.

The APS is, as already mentioned, based on a cylindrical resonator. The plasma is confined in a quartz tube, since quartz is transparent for a wide range of electromagnetic radiation. Thus a wide optical range can be used for the characterisation by optical emission spectroscopy of the plasma. In further industrial applications the quartz tube can be replaced by a tube made of a more robust material, which only has to fulfil the requirement that it is transparent for microwaves, for example by a ceramic tube. The gas is supplied via a metallic nozzle, which forms a coaxial structure below the cylindrical resonator and is also needed as a tuning element to adjust the resonant frequency, as explained in chapter 3. Additionally to the central gas inlet via the nozzle, a second tangential

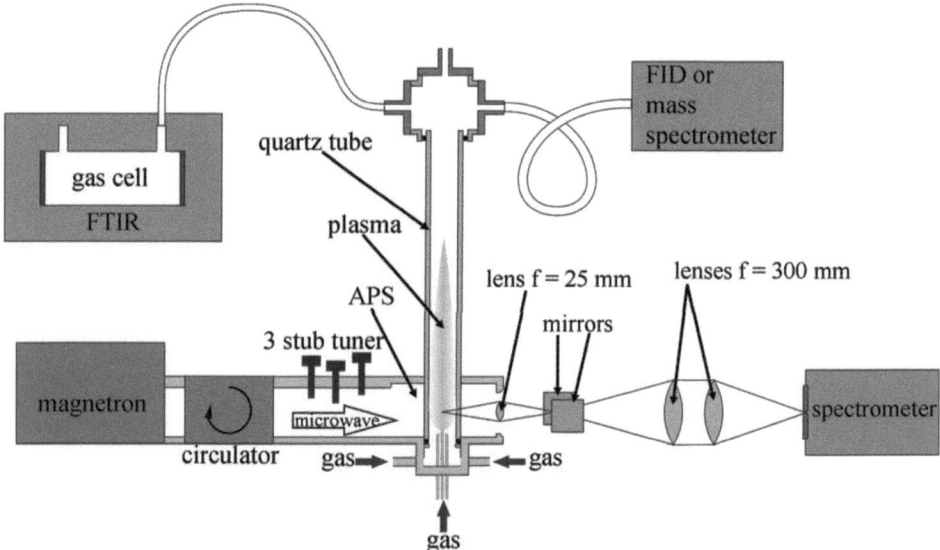

Figure 4.1: Schematic view of the experimental setup, which was used to characterise the plasma of the APS and for the analyses of the decomposition of waste gases. The characterisation of the APS plasma was performed by means of optical emission spectroscopy. The plasma is imaged by the optical setup consisting of three lenses (one $f = 25\,\text{mm}$ and two $f = 300\,\text{mm}$) and two mirrors on the entrance slit of a spectrometer. The analyses of the raw and clean gases were performed with the shown Fourier-Transform Infra-Red spectrometer (FTIR), a flame ionisation detector (FID), and a quadrupole mass spectrometer.

gas inlet was installed. The tangential gas inlet leads to a vortex of the plasma gas, so that the plasma is kept away from the quartz tube and so the plasma is stabilised. Additionally, the vortex leads to a better mixing of the gas, which supports the exhaustive decomposition of waste gases. The microwave of a frequency of 2.45 GHz is supplied via rectangular waveguides and the circulator from the magnetron to the APS, which was explained in detail in section 3.2. In the following two sections the experimental setup, which is needed for the optical emission spectroscopy, and the measurement devices for the analyses of the decomposition of waste gases, will be presented in detail.

4.1.1 Optical Emission Spectroscopy

The characterisation of the APS plasma is performed by means of optical emission spectroscopy since it has the following advantages:

- Almost all temperatures and densities of all plasma particles can be determined.

4.1. THE EXPERIMENTAL SETUP

- It is a non disturbing diagnostic, since exclusively the voluntarily emitted radiation from the plasma is utilized.

- Since, the diagnostic is in no contact with the plasma, reactive plasmas, like plasmas containing waste gas, can be characterised.

To characterise the plasma inside the resonator, the front side of the APS was furnished with a narrow slit of $d = 4$ mm, which does not affect the electrical field distribution in the resonator.

Three different spectrometers were used for the characterisation: two overview spectrometers, which are sensitive in the visible and IR region 200..1000 nm (Mechelle 7500) and in the UV and visible region 175..700 nm (Avantes), respectively, and furthermore a third spectrometer, which can also be used to measure overview spectra as well as for the measurement of spectra with a high resolution in the wavelength range of 200 nm..800 nm (Acton, SpectraPro-750i). The Mechelle spectrometer has a focal length of 190 mm and a resolution of $\lambda/\Delta\lambda = 7500$. The Avantes spectrometer has a resolution of about $\lambda/\Delta\lambda \approx 1000..3500$. The Acton spectrometer is equipped with three different gratings: 150, 600 and 1800 grooves mm^{-1}. With the 150 grooves mm^{-1} grating overview spectra in the UV and visible region can be measured while the 1800 grooves mm^{-1} grating is used for the measurement of high resolution spectra. The Acton spectrometer has a focal length of 750 mm and therefore a maximal resolution of $\lambda/\Delta\lambda = 20000$ can be achieved with the 1800 grooves mm^{-1} grating. An image-amplified CCD camera with 1300 × 1030 pixel is mounted at the exit slit of the Acton spectrometer. This camera reduces the maximal resolution to $\lambda/\Delta\lambda = 13000$ but has the advantage that simultaneously to the spectral measurement a one-dimensional spatial resolution is obtained. Table 4.1 summarises the properties of the three utilised spectrometers.

For a spatially resolved characterisation of the plasma in radial and axial direction, the plasma was projected on the entrance slit of one of the three spectrometers by an optical setup. A schematic view of the optical setup is shown in Fig. 4.1. For a complete radial characterisation of the plasma inside the resonator, a lens with a focal length of $f = 25$ mm was placed very close to the slit in the resonator. The two mirrors were used to turn the image of the plasma by 90°. Then two lenses with focal lengths of $f = 300$ mm are used to sharply project the virtual intermediate image of the plasma on the entrance slit of the spectrometer. All components of the optical setup were chosen to have good optical properties from the UV to the IR range. When the Acton spectrometer was used, a one-dimensional spatial resolution in radial direction was already achieved by the CCD camera. When the other spectrometers were used the spectrometer was moved along the radial direction. To get an axial resolution of the plasma the whole optical setup and the spectrometers were moved in axial direction while the APS remained at the same position. At each axial position the optical setup was checked and if necessary readjusted. Fig. 4.2 illustrates the optical setup by

Table 4.1: Overview of the three utilised spectrometers.

	Avantes	Mechelle	Acton
resolution $\lambda/\Delta\lambda$	$\approx 1000..3500$	7500	13000
observance wavelength	175..700 nm high sensitivity in the UV and visible range	200..1000 nm high sensitivity in the visible and IR range	200..800 nm high sensitivity in the visible range
advantages	• small and portable • easy relative intensity calibration	overview spectrometer with a high resolution	• three gratings 150, 600 and 1800 grooves mm^{-1} • one-dimensional, spatial resolution

photos.

The properties of the optical setup and the magnification factor were measured with a 1 mm-slit which was located in the centre of the APS, was illuminated with different lamps for the analysed wavelength ranges and moved in radial direction in the APS. Fig. 4.3 a) shows a photo of a HgCd-lamp, which illuminates the 1 mm-slit. The quartz tube is placed in front of the slit. The lens holder with the $f = 25$ mm lens can be seen on the left side of the photo and the Acton spectrometer in the background. The two mirrors and the other two lenses with $f = 300$ mm are not shown in the photo but of course were also part of the optical setup as illustrated in Fig. 4.2. A diagram of the transformation properties of the optical setup on the camera of the Acton spectrometer for a wavelength of 313 nm (Hg spectral line) is shown in Fig. 4.3 b) as an example. Such diagrams were measured for all analysed wavelength ranges and than used to reconstruct the spatial distribution of the plasma parameters in the APS.

Since absolute line intensities or at least line intensity ratios are necessary for the determination of the plasma parameters, like particle densities and temperatures, an intensity calibration of the utilized spectrometers was performed with a tungsten ribbon-lamp. The tungsten ribbon-lamp was placed in the centre of the APS, comparable to the HgCd-lamp in Fig. 4.3 but without the slit, for this purpose. Then a spectrum of the tungsten ribbon-lamp was recorded and by comparison with the theoretical spectrum of the tungsten ribbon-lamp an intensity calibration function was calculated. This calibration function was used to correct the measured intensities of the observed transitions.

4.1. THE EXPERIMENTAL SETUP

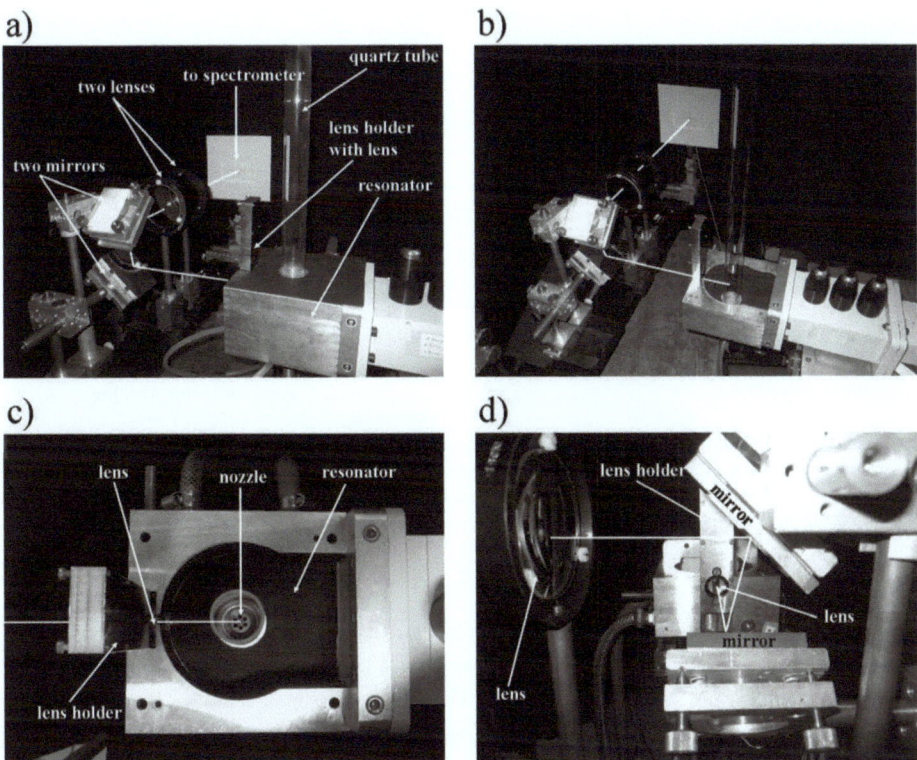

Figure 4.2: Photos of the optical setup, which was used to characterise the plasma by means of optical emission spectroscopy. a) and b) show the whole optical setup in a) with the closed resonator and b) with the opened resonator. c) Shows a close look from above at the opened resonator with the metallic nozzle in its centre and at the lens holder of the $f = 25$ mm-lens. d) Shows a front view of the opened resonator with the lens holder of the $f = 25$ mm-lens, the $f = 25$ mm-lens, and the two mirrors. The optical path is drawn in white in all the photos.

4.1.2 Decomposition of Waste Gases

One possible application of the APS could be the decomposition of waste gases. To examine if the APS is suitable for the abatement of waste gases, the decomposition of volatile organic compounds (VOC) and perfluorinated compounds (PFC) was studied in this work. The raw and clean gases were analysed with a Fourier-Transform Infra-Red spectrometer (FTIR) from Bruker Vector 22 with a gas cell with a length of 20 cm, a flame ionisation detector (FID), a quadrupole mass spectrometer, and an Agilent GC/MSD 6890/5970 gas-phase chromatograph.

The raw and clean gases are guided through the FTIR, the FID and/or the quadrupole mass spectrometer after they have passed the APS as shown in Fig. 4.1. A sample of the clean gas was absorbed on Tenax® and then analysed in the gas-phase chromatograph. The desorption of the

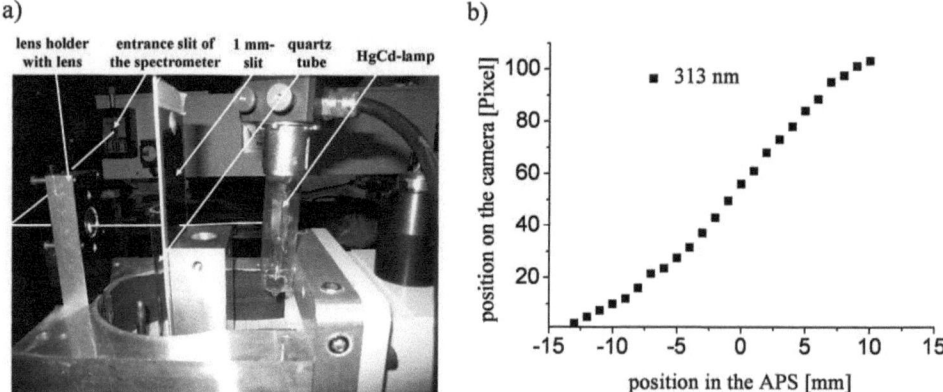

Figure 4.3: a) Shows a photo of the optical setup which was used to measure the properties of the optical setup and its magnification factor. A 1 mm-slit, which is placed in the centre of the APS, is illuminated with a lamp, here a HgCd-lamp, and is moved in radial direction. b) Shows a diagram of the transformation properties of the optical setup on the camera of the Acton spectrometer for a wavelength of 313 nm (Hg spectral line).

sample occurred thermally in a helium gas flow with a Gerstel TDS 2-system.

The raw and clean gas analyses enabled to get knowledge about the generated products. The results from the analyses performed with the FID and the quadrupole mass spectrometer were also used to calculate destruction and removal efficiencies DRE of the waste gases. Furthermore, optical emission spectroscopy of plasmas containing waste gas was performed to obtain information about the reaction products in the plasma. The detailed characterisation of the raw and clean gases and the spectroscopic analyses of the plasmas containing waste gas led to information about the possible reaction channels. All these results of the characterisation of the decomposition of waste gases with the APS are presented in section 4.3.

4.2 Characterisation of the APS Plasma

One possible application of the APS is the abatement of waste gases, which is performed in air or nitrogen plasmas, since usually waste gases appear as mixtures of air or nitrogen and the harmful gases. For a comprehensive understanding of the decomposition of waste gases and to influence possible reaction channels as well as to optimise the decomposition process, detailed information about the particle densities and temperatures in dependence of the gas flow and microwave power of these plasmas is required. Therefore, the plasma is characterised by means of optical emission spectroscopy which uses the voluntarily emitted radiation of the plasma.

Since the radiation of the plasma is determined by the densities and the temperatures but also by the quantum mechanical structure of the molecules and atoms these must be known to get distinct and reliable information about the temperatures and densities. Therefore, the following section

4.2.1 covers the theoretical description of atomic and molecular spectra.

Thereafter, overview spectra of an air plasma are presented and it is discussed which plasma parameters are available by evaluating these spectra and which spectroscopic methods and evaluation procedures are chosen for this purpose. This is presented in section 4.2.2.

Finally the gas and electron temperatures as well as their densities for an air plasma for different gas flows and microwave powers are presented in section 4.2.3. Furthermore, these results are discussed and evaluated as well as considerations about the heating mechanism are shown in this section.

4.2.1 Theoretical Descriptions of Atomic and Molecular Spectra

Since usually optical emission spectra of atmospheric microwave plasmas exhibit radiation from molecules as well as from atoms the following sections introduce the spectra of atoms and diatomic molecules which are presented in section 4.2.1.1 and 4.2.1.2, respectively. It is also shown what kind of temperatures and densities can be determined in each case.

4.2.1.1 Spectra of Atoms

The spectrum of an atom consists of individual spectral lines. This can already be descriptively explained with a very simple description of atoms, which is provided by Bohr's atomic model. According to Bohr's atomic model, an atom consists of a positively charged atomic nucleus and electrons, which circulate around the nucleus on discrete orbits, like planets around the sun. The radiation from an atom in form of a photon is emitted when an electron from a higher orbit with a higher energy falls down to an orbit with a lower energy, not necessarily the ground state.

If the radiation emitted by atoms in a plasma shall be used to gain information about particle densities and temperatures, a more detailed description to calculate the atomic transitions is indispensable. A proper description is provided by quantum mechanics. The electrons are no longer regarded as particles but as waves. Thus the electrons do not circulate in discrete orbits around the nucleus and their residence is not exactly designated. The solutions of Schrödinger's equation for electrons in the electromagnetic field of the nucleus lead to probabilities of presence for the electrons, which correspond to different energy levels [56]. Some of these energy levels can be degenerated and therefore can be populated by more electrons.

At the absolute zero point of temperature the electrons populate the lowest energy levels with respect to the Pauli principle since electrons are fermions [56]. The higher energy levels are populated either by inelastic collisions with other particles or by the absorption of a photon, which is called collisional or photo excitation, respectively. The deexcitation can occur in two different ways: By a collision with another particle, which is called collisional deexcitation, or by the emission of a photon. The emission of a photon can either be induced by another photon with the same energy or happens spontaneous and is then called induced or spontaneous emission, respectively. However, not all energetically possible transitions between two levels are allowed. Certain selection

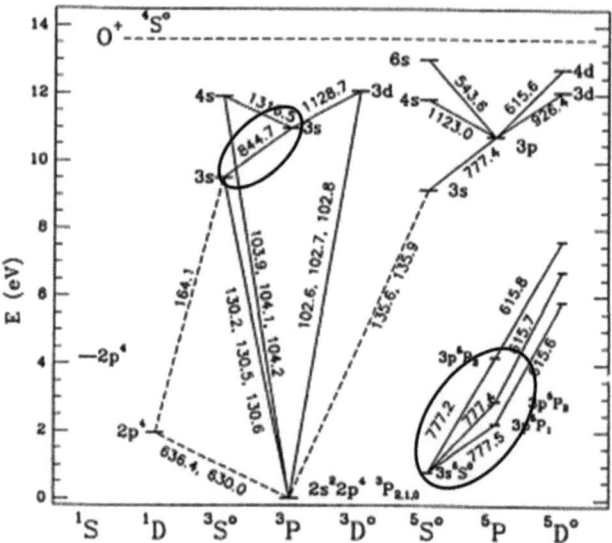

Figure 4.4: Grotrian diagram for some energy levels of oxygen from [59]. ^{1}S, ^{1}D, ^{3}S^0, and so on are the atomic energy states. The transitions $^5P \rightarrow {}^5S^0$ around 777 nm and $^3P \rightarrow {}^3S^0$ around 844 nm, which are marked with the ellipses, are observed in an APS air plasma and are utilised to determine the excitation temperature T_{ex} by means of a Boltzmann plot as will be explained in section 4.2.2.

rules specify which transitions are allowed and are for example presented in [56].

The energy levels, which are also called the spectral series, can be represented graphically by a Grotrian diagram. A Grotrian diagram for some energy levels of oxygen is shown in Fig. 4.4 including the transitions and their wavelength.

The energy difference ΔE of the two participating energy levels defines the frequency ν by $\Delta E = h\nu$ and the wavelength $\lambda = \frac{c}{\nu}$ of the emitted photon. Thus this radiation should be a very sharp line, but the spectral line can be broadened by different mechanisms.
First of all, all spectral lines have a natural line width, which is caused by the fact that no exact energy can be assigned to the energy levels due to Heisenberg's uncertainty principle. Since the two energy levels have no distinct energy, the energy of the emitted photon also varies some what, which causes the natural width of the spectral line.
Furthermore, the spectral line can be broadened by an electric field, for example induced by the electrons, which causes a Stark broadening. The interaction with other particles causes a further broadening, the Van der Waals and pressure or collisional broadening. Moreover, the movement of the atom itself leads to a Doppler broadening of the spectral line. Lastly, the broadening of the spectral line by the optical setup must be considered, which is called apparatus broadening.

4.2. CHARACTERISATION OF THE APS PLASMA

The different line broadening mechanisms can be used to determine some temperatures and densities: Since the Doppler broadening is caused by the movement of the atoms, the Doppler broadening provides information about their translation temperature T_g. The highest broadening can be observed with light atoms, such as hydrogen.

The Stark broadening can be used to determine the electron density. Commonly also transitions of the hydrogen atoms are used, especially the H_β atomic line, since the linear Stark effect can be applied to this transition and well-engineered evaluation methods have been developed.

The intensity of a spectral line I_{ik} is given by the following relation, if a Boltzmann distribution is present:

$$I_{ik} = \frac{h\nu_{ik}}{4\pi} A_{ik} \frac{g_k}{Q_k} n_0 e^{-\frac{E_k}{k_b T}}, \qquad (4.1)$$

where ν_{ik} is the frequency of the spectral line, g_k is the statistical weight of the upper level, Q_k is the partition function of the upper level, E_k is the energy of the upper level, n_0 is the number of atoms, and A_{ik} is Einstein's transition probability. The transition probability A_{ik} can be calculated by quantum mechanics by the overlapping of the wave functions of the two participating levels i and k. The A_{ik}s have been calculated for many kinds of atoms and are tabulated, so that the values can be looked up. Thus if two spectral lines with different excitation energies E_k and E_l are observed, the temperature can be determined by the following equation:

$$T = -\frac{(E_k - E_l)}{k_b \cdot ln\left(\frac{I_{ik}\nu_{jl}A_{jl}g_l}{I_{jl}\nu_{ik}A_{ik}g_k}\right)} \qquad (4.2)$$

whereas ik and jl refer to the two participating transitions. If more than two spectral lines are observed a Boltzmann plot leads to a more precise determination of the electron temperature. Then the logarithm of the reduced intensity $ln\left(\frac{I_{ik}}{\nu_{ik}A_{ik}g_k}\right)$ is plotted against the energy of the upper level E_k and a temperature T can be calculated from the slope $\frac{E_k}{k_b T}$ of the obtained straight line. The obtained temperature is called excitation temperature T_{ex}.

In local thermodynamic equilibrium the excitation temperature T_{ex} is the electron temperature T_e. Even in partial local thermodynamic equilibrium, provided that the electron density is high enough, the excited levels and the free electrons and the ions are balanced with only the ground state differing from local thermodynamic equilibrium. Then the Boltzmann plot is still valid to determine the electron temperature T_e [48, 47].

Equation 4.1 is not only dependent on the temperature but also on the number of atoms n_0. The density of different atomic species can therefore be determined from the absolute value of the spectral line intensity, when all quantum mechanical constants as well as the temperature are known.

To summarise, the spectra of atoms can be used to determine the electron temperature T_e, the gas temperature T_g and atom densities as well as the electron density n_e. The electron temperature can be deduced from the line intensity ratios by using a Boltzmann plot while the density of different atomic species can be determined from the absolute value of the line intensity. The gas temperature and the electron density can be deduced from the line broadening caused by the Doppler effect and by the Stark effect, respectively.

4.2.1.2 Spectra of Diatomic Molecules

Generally, in a simple model molecules are built of several nuclei with electrons around them. Comparable to the atoms, described in the previous section 4.2.1.1, molecules have electronic energy levels. The electrons occupy the different levels and the molecule can be excited and deexcited electronically. Furthermore, since molecules consist of more than one nucleus, molecules not only have electronic energy levels: The whole molecule can rotate and the single nucleus in the molecule can oscillate against each other. The resulting rotational, vibrational, and electronic energy levels can be regarded separately, since the particular energies are distinctly different: $E_{rot} < E_{vib} < E_{el}$.

In the present work only the spectra of diatomic molecules, which are the simplest molecules, will be explained and discussed, since primarily spectra of diatomic molecules are observed in the analysed APS plasma.

As a first approximation the rotation of a diatomic molecule can be described by the handle model of a rigid rotator [57]. The energy of a certain level is discrete and can be calculated by Schrödinger's equation to:

$$E_{rot} = \frac{h^2}{8\pi^2 I} J(J+1), \qquad (4.3)$$

where h is Planck's constant, I is the momentum of inertia, and J is the rotational quantum number. The rotational quantum number J can be an integer $0, 1, 2, ..$ and is correlated to the angular momentum $\vec{L}$ by the following equation:

$$|\vec{L}| = \frac{h}{2\pi} \sqrt{J(J+1)}. \qquad (4.4)$$

In molecular spectroscopy a common unit for the energy of the emitted photon is cm^{-1} and the rotational energy becomes:

$$F(J) = \frac{1}{\lambda} = \frac{E_{rot}}{hc} = BJ(J+1) \qquad (4.5)$$

with

$$B = \frac{h}{8\pi^2 cI}, \qquad (4.6)$$

which is called the rotational constant B. Since the diatomic molecule is not completely rigid and thus the distance between the two nuclei varies in dependence of the angular momentum due to the centrifugal force, a correction term must be added in equation 4.5:

$$F(J) = BJ(J+1) - DJ^2(J+1)^2, \qquad (4.7)$$

where D is approximately given by the Kartzer relation:

$$D = \frac{4B^3}{\omega^2} \qquad (4.8)$$

4.2. CHARACTERISATION OF THE APS PLASMA

and is always small compared to B.

A Morse potential is assumed for the description of the oscillation of the molecule [57]:

$$V = D_e \left(1 - e^{-\beta(r-r_e)}\right)^2, \tag{4.9}$$

where D_e is the dissociation energy, r_e is the potential minimum, and β is given by:

$$\beta = \sqrt{\frac{2\pi^2 c \mu_A}{D_e h}} \omega_e. \tag{4.10}$$

μ_A is the reduced mass of the molecule and ω_e is the oscillation frequency in cm^{-1}. The oscillation energy can again be calculated by Schrödringer's equation and one obtains:

$$G(\nu) = \omega_e \left(\nu + \frac{1}{2}\right) - \omega_e x_e \left(\nu + \frac{1}{2}\right)^2 + \omega_e y_e \left(\nu + \frac{1}{2}\right)^3 + ..., \tag{4.11}$$

where ν is the oscillation quantum number. The higher-order correction terms are much smaller compared to the lower-order ones. Usually the oscillation and the rotation superpose and therefore, both must be combined. Therefore, the constant B, which is dependent on the distance of the two nuclei, is split into a part which describes the equilibrium position B_e and another oscillation dependent part B_ν. This leads to:

$$B_\nu = B_e - \alpha_e \left(\nu + \frac{1}{2}\right) + ... \tag{4.12}$$

with

$$B_e = \frac{h}{8\pi^2 c \mu_A r_e^2}. \tag{4.13}$$

Compared to B_e the constant α_e is small, which depends on the anharmonicity of the oscillation and on B_e and ω_e. Furthermore, like the oscillation is taken into account for B, D must also be corrected:

$$D_\nu = D_e - \beta_e \left(\nu + \frac{1}{2}\right) + ... \tag{4.14}$$

When all the above results are combined, the rotational vibrational energy term becomes:

$$F(J) = B_\nu J(J+1) - D_\nu J^2 (J+1)^2, \tag{4.15}$$

where the index ν denotes the examined oscillation level.

As mentioned above an electronic energy term E_{el} exists alongside the rotational and vibrational energy term, so that the total energy of an energetic level of a molecule is given by:

$$E_{tot} = T_e + G(\nu) + F_\nu(J) \tag{4.16}$$

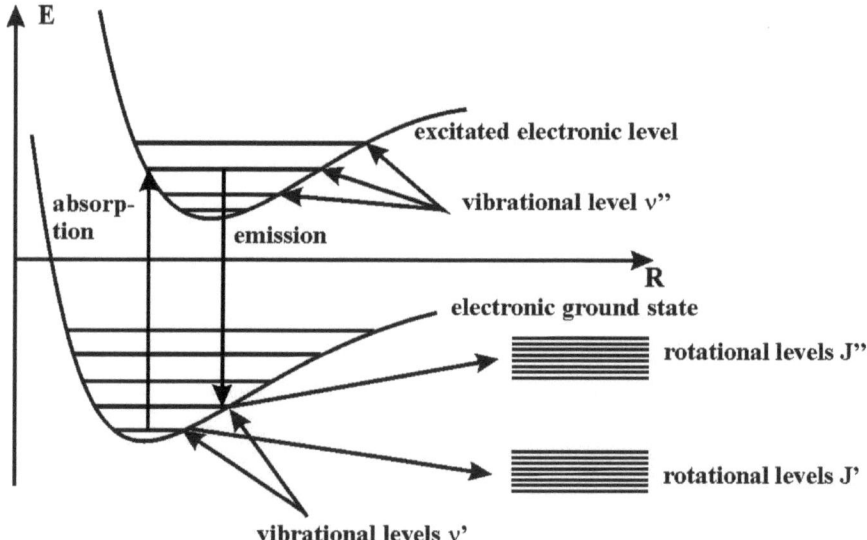

Figure 4.5: Schematic view of the energy levels of a molecule. The electronic levels are subdivided by vibrational levels, which are again subdivided by rotational levels.

where T_e denotes the contribution from the electronic excitation. Thus the energy of an energy level of a molecule has a rotational, vibrational and electronic amount which is illustrated in the schematic view in Fig. 4.5. The electronic levels are subdivided by vibrational levels, which are again subdivided by rotational levels.

Like the excitation of atoms the population of the energy levels is provided either by collisional excitation with other particles or by the absorption of a photon. The deexcitation is also again possible by collisional deexcitation or by induced or spontaneous emission of a photon. For the transition between the different energetic levels again certain transition rules must be satisfied, which are for example explained in [57].

If the population of the energetic levels follows a Boltzmann distribution, the intensity of a spectral line is given by:

$$I_{ik} = \frac{h\nu_{ik}}{4\pi} A_{ik} \frac{g_k}{Q_k} n_0 e^{-\frac{E_k}{k_b T}}, \qquad (4.17)$$

where n_0 is the number of molecules, g_k is the energetic degradation of the upper energy level, E_k is the energy of the upper energetic level, ν_{ik} is the frequency of the emitted photon, Q_k is the partition function of the upper level, h and k_b are Planck's and Boltzmann's constant, respectively, and A_{ik} is the transition probability. The transition probabilities can again be calculated by quantum mechanics from the overlapping of the wave functions of the participating energetic levels. Since the intensity of spectral lines of molecules is dependent on the temperature the spectra of molecules can be used to determine temperatures of particles in the plasma.

Pure rotational transitions are observed in the wavelength range of microwaves while pure vi-

4.2. CHARACTERISATION OF THE APS PLASMA

brational and rotational-vibrational transitions are observed in the IR range. Pure electronic transitions or electronic transitions in combination with changes in the rotational and/or vibrational quantum number are located in the visible and UV wavelength range.

Since optical emission spectroscopy usually covers the wavelength range between 150 nm and 200 nm to 1 μm, only electronic transitions of molecules are observed. Since electronic transitions with no change of the rotational and vibrational quantum number are possible, but also electronic transitions where the rotational and/or the vibrational quantum number changes, the spectra of molecules are very complex but therefore also offer the possibility to gain much information about the plasma.

For electronic transitions where only the rotational quantum number J changes and the vibrational quantum number ν stays unchanged the temperature in equation 4.17 provides information about the rotational temperature T_{rot} of the regarded molecule. Since usually the energy transfer between the rotational energy of a molecule and its translation energy in atmospheric pressure plasmas is warranted, the rotational temperature provides a good estimation of the gas temperature of the plasma.

When electronic transitions are regarded where only the vibrational quantum number ν changes, information about the vibrational excitation of the molecule is given, which is expressed by the vibrational temperature T_{vib}. Normally the vibrational temperature T_{vib} is higher compared to the rotational temperature T_{rot} but still smaller than the electron temperature T_e.

Furthermore, it must be noted that the spectral lines are also broadened by the same effects as the spectral lines of atoms, so that the translation or gas temperature T_g can also be determined from the Doppler broadening of the molecular spectral lines.

Thus the spectrum of a molecule provides information about the temperatures of the molecule. When different molecule species are present in the plasma the particular spectra of each species give information about the particular temperatures of the different plasma particles. Since furthermore different temperatures T_g, T_{rot}, and T_{vib} can be determined for each species from the molecular spectra, it can be analysed how well the energy is transferred within the molecules and between the different species and therefore, to what extend the plasma is in (partial) local thermodynamic equilibrium.

Since the intensity of the spectral lines is also dependent on the number of molecules, the density of each species can be determined from the absolute value of the line intensity when all quantum mechanical constants for the regarded transitions and the temperature determined from line intensity ratio considerations are known.

The spectra of molecules provide information about different temperatures and densities of the species present in the plasma. Rotational and vibrational temperatures, T_{rot} and T_{vib}, can be deduced from line intensity ratios while the translation temperature T_g can be determined from

Table 4.2: Summary of possible methods to determine the fundamental plasma parameters by optical emission spectroscopy.

spectroscopic method	plasma parameter	applied for the characterisation of the APS plasma
atomic spectra		
line intensity ratios: Boltzmann plot	excitation or electron temperature T_{ex} or T_e	✓
absolute line intensity	density of atomic species	✗
line width: Doppler broadening	translation temperature T_g of the atoms	✗
Stark broadening	electron density n_e	✗
molecular spectra		
line intensity ratios	rotational and vibrational temperature, T_{rot} and T_{vib}	✓
absolute line intensity	density of molecular species	✗
line width: Doppler broadening	translation temperature T_g of the molecules	✗

the line broadening caused by the Doppler effect. The absolute intensity of the spectral lines can be used to determine the densities of the different molecule species.

To summarise, the detailed description of atom and diatomic molecule spectra, can be used to obtain detailed information about fundamental plasma parameters: particle temperatures and densities. Table 4.2 summarises the presented methods to determine these plasma parameters by optical emission spectroscopy. It shows, that line intensity ratios of spectral lines from atoms and molecules can be used to measure particle temperatures if the energy levels are Boltzmann populated, while the absolute line intensities provide information about the particle densities. Furthermore, if the line width is dominated by Doppler or Stark broadening mechanisms, the line width can be used to determine either the translation temperature T_g or the electron density n_e, respectively.

The right column in table 4.2 denotes which spectroscopic methods were applied to measure the fundamental plasma parameters of the APS plasma. Some methods could not be applied since the necessary transitions could not be observed, other plasma parameters were determined by using

4.2. CHARACTERISATION OF THE APS PLASMA

another method. A detailed description why these methods were used and which other methods were chosen to measure the plasma parameters of the APS plasma is presented in the following section 4.2.2.

4.2.2 Overview Spectra and Applied Spectroscopic Methods

To acquire preliminary information about the APS plasma and to obtain knowledge which spectroscopic methods can be applied to determine the plasma parameters overview spectra of an air plasma were recorded. Fig. 4.6a) shows overview spectra of dry and humid air plasma in the UV and visible range from 175 mm to 725 mm recorded by the Avantes spectrometer.

The spectrum of the dry air plasma is dominated by molecule NO-bands in the UV. These bands belong to the $B^2\Pi - X^2\Pi$- and $A^2\Sigma^+ - x^2\Pi$-transitions, which are called the NO$_\beta$- and NO$_\gamma$-system, respectively [58]. These NO-bands outshine N_2- and N_2^+-bands as well as O_2-bands, which therefore are not observed. When the air is moistened, additionally to the NO-bands, molecule bands of the free OH radical appear in the UV. These molecule bands can be used to determine rotational and vibrational temperatures of the particles as explained in section 4.2.1.2. Since the energy transfer between the rotational excitation of a molecule and its translation energy is usually ensured, the rotational temperature provides a good estimate of the gas temperature. The $A^2\Sigma^+ - X^2\Pi_\gamma$-transition of the free OH radical between 306 nm and 310 nm is very sensitive to the temperature and is therefore often used to determine the rotational temperature T_{rot} [60, 61, 62]. Furthermore, studies performed by E. Felizardo et al. showed that the admixture of water as a diagnostic gas only has a negligible influence on the gas temperature and therefore can well be used for the determination of the gas temperature [65].

A method which allows relatively quick and easy an estimation of the rotational temperature was proposed by C. Izarra and is based on the comparison of three characteristic band heads at 306.50 nm, 306.91 nm, and 309.50 nm [61]. A more precise method to measure the rotational temperature was developed by J. Happold and is based on the comparison of a measured high-resolution spectrum of the free OH radical between 306 nm to 309.5 nm with simulated spectra of different temperatures [62, 63]. The simulation of the spectra is based on quantum mechanical calculations and on high-resolution measurements by Dieke and Crosswhite [64]. For the population of the particular levels a Boltzmann population was assumed [62, 63]. The spectra were convolved with the apparatus profile with an apparatus width of $\Delta = 0.06$ nm. Since the intensity ratios of the bands already provide the information about the rotational temperature, all spectra were normalised to the band head at 308.9 nm.

Fig 4.7 a) shows simulated OH-spectra for different temperatures between 830 K and 4440 K. It can easily be seen that the shape of the spectrum changes distinctly when the temperature is varied and therefore is suitable to determine the rotational temperature.

The high-resolution spectra were measured with the Acton spectrometer with the 1800 grooves · mm^{-1} grating. Since the Acton spectrometer is equipped with an ICCD camera simultaneously to

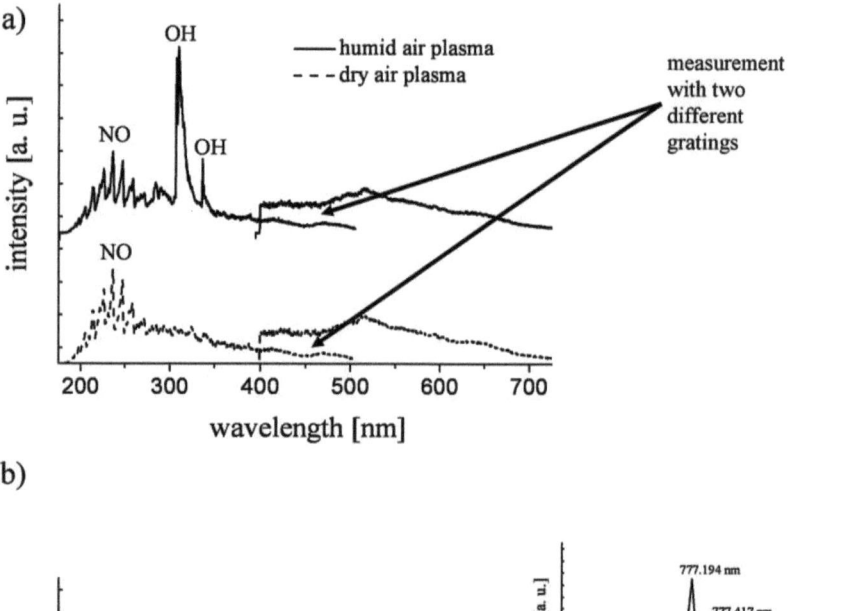

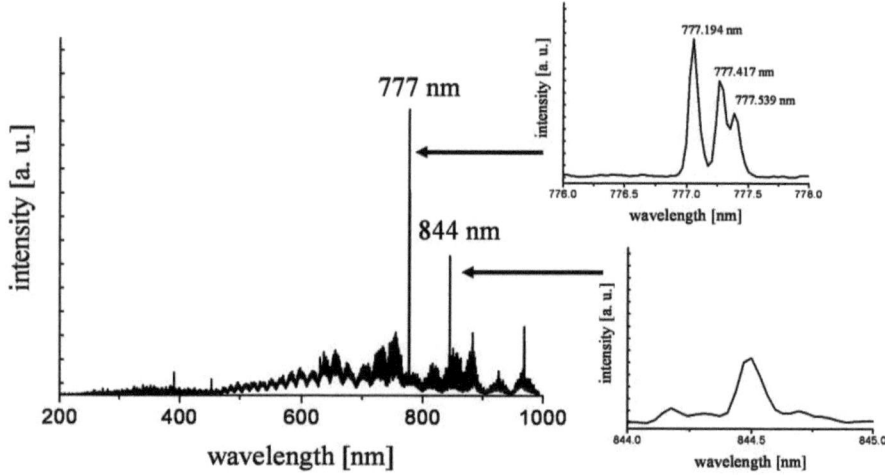

Figure 4.6: Overview spectra of an air plasma: a) Overview spectrum in the UV and visible range of a dry and humid air plasma recorded with the Avantes spectrometer. Since the Avantes spectrometer is equipped with two different grating, one for the wavelength range of 175 nm..500 nm and the other of 400 nm..700 nm, in the wavelength range between 400 nm and 500 nm measurements with both gratings are conducted which is why here two spectra with different intensities are shown. The spectrum of the dry air plasma is dominated by NO-bands in the UV while in the spectrum of the humid air plasma supplemental OH-bands appear. b) Overview spectrum from the UV to the IR range of a dry air plasma recorded with the Mechelle spectrometer. The sensitivity of the Mechelle spectrometer is very poor in the UV range and therefore the NO-bands can only be observed weakly. However, with the Mechelle spectrometer the IR range is accessibly, where two atomic oxygen lines are observed at 777 nm and 844 nm.

4.2. CHARACTERISATION OF THE APS PLASMA

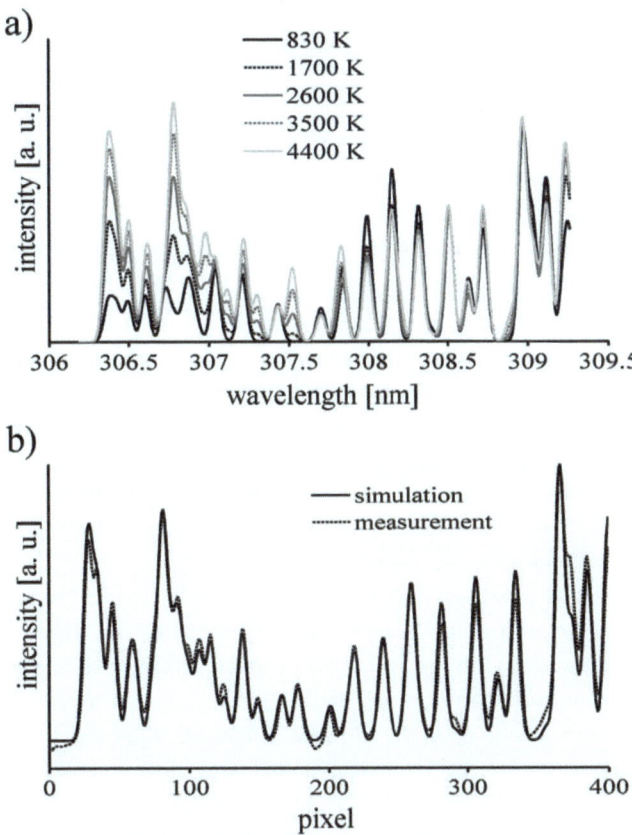

Figure 4.7: a) simulated spectra of the $A^2\Sigma^+ - X^2\Pi_\gamma$-transition of the free OH radical between 306 nm and 309.5 mm for different temperatures between 830 K and 4400 K. b) Comparison of a measured spectrum to the simulated spectrum which agrees best with the measured one. Thus the temperature could be determined to be 3140 K. Since already the intensity ratios provide the information about the rotational temperature, all spectra are normalised to the band head at 308.9 nm.

the spectral measurement a one-dimensional spatial resolution is obtained. Thus in combination with the two mirrors, which turn the image by 90°, a profile in radial direction could be measured simultaneously.

Fig. 4.6 shows that the spectrum of a dry air plasma already exhibits NO-bands where the OH-band, which is used to obtain the rotational temperature, is located. Thus the high-resolution spectra of the OH-band must be corrected. Therefore, for each parameter set a spectrum of a dry and humid air plasma was recorded between 306 nm..310 nm. Then the spectrum of the dry air plasma was subtracted from the spectrum of the humid air plasma. Since only intensity ratios are needed to determine the rotational temperature, the measured spectra were also normalised to the band head at 308.9 nm. Furthermore, only a very narrow wavelength range between 306 nm and 309.5 nm was used and therefore, an intensity calibration was not necessary for the evaluation of these spectra.

To obtain the rotational temperature from the measured and corrected spectra, spectra for temperatures from 830 K to 9800 K were simulated in 30 K steps. The measured spectra were compared to the simulated spectra and then the temperature with the best agreement of the two spectra was taken [62, 63]. Fig. 4.7 b) shows a measured high-resolution spectrum and the simulated spectrum, which fits best with the measured one. In this case the temperature could be determined to $T_{rot} = 3140$ K.

An overview spectrum from 200 nm to 1000 nm of a dry air plasma is shown in Fig. 4.6 b). This spectrum was recorded by the Mechelle spectrometer. Since the sensitivity of the Mechelle spectrometer is very poor in the UV range, the NO-bands, which were observed clearly with the Avantes spectrometer, cannot be seen very well in this overview spectrum. However, with the Mechelle spectrometer the IR range is accessible, which revealed that two atomic oxygen lines at 777 nm and at 844 nm can be observed. Atomic nitrogen lines which are located around 925 nm, 850 nm..875 nm and 900 nm were not observed.

These two atomic oxygen transitions $^5P \rightarrow {}^5S^0$ at 777 nm and $^3P \rightarrow {}^3S^0$ at 844 nm were used to obtain the excitation temperature T_{ex} by means of a Boltzmann plot. T_{ex} gives a lower estimation of the electron temperature T_e. All spectra of the two atomic oxygen lines for the determination of the electron temperature were also recorded by the Mechelle spectrometer, since both other spectrometers have a too low sensitivity in the IR range. Unlike the Acton spectrometer, the Mechelle spectrometer has no one-dimensional resolution, so that many single spectra had to be recorded for the measurement of the radial profiles. The line intensity ratios are needed for the determination of the electron temperature with a Boltzmann plot. Since the two observed atomic oxygen lines are located in a wide wavelength range, an intensity calibration is necessary. Therefore, the Mechelle spectrometer was calibrated with a tungsten ribbon-lamp as explained in section 4.1.1.

Furthermore, since the refraction power of the lenses is dependent of the wavelength the optical setup had to be adjusted for the measurement of each of the two different temperatures.

Thus the overview spectra of a dry and a humid plasma showed that the $A^2\Sigma^+ - X^2\Pi_\gamma$-transition of the free OH radical between 306 and 310 mm is observed in humid air plasmas and can be used to obtain the rotational temperature T_{rot}, which gives a good estimate of the gas temperature T_g. Furthermore, the overview spectrum shows two atomic oxygen lines at 777 nm and at 844 nm, which can be used to determine the excitation temperature T_{ex} by means of a Boltzmann plot. T_{ex} provides a lower estimation of the electron temperature T_e.

However, the neutral particle and electron density, n_a and n_e, cannot be obtained directly from these spectra.

A well established method to determine the electron density n_e is measuring the Stark broadening of the H_β atomic line at 486 nm, as explained in section 4.2.1.1. However, due to the moisture of the air, only OH-bands but no atomic hydrogen lines could be observed, but since the H_β atomic line offers the possibility to measure the electron density, little amounts of pure hydrogen were mixed to the air as diagnosis gas. Nevertheless only one atomic hydrogen line, the H_α line at 656 nm could be observed. This line is however predominantly broadened by the Doppler effect and therefore cannot be used to determine the electron density. Because of that, the electron density could not be measured directly by optical emission spectroscopy.

The neutral particle density n_a can be estimated by using Dalton's law 1.10:

$$p = n_e k_b T_e + (n_a + n_i) k_b T_g.$$

Assuming $n_i = n_e \ll n_a$, n_e and n_i are negligible against n_a and so equation 1.10 becomes:

$$p = n_a k_b T_g. \tag{4.18}$$

Rearranging this equation leads to:

$$n_a = \frac{p}{k_b T_g}. \tag{4.19}$$

The neutral particle density can thus be calculated from the measured gas temperature, if p is assumed to be constant and equal to the surrounding atmospheric pressure.

Since the electron density could not be measured directly by optical emission spectroscopy, the electron density was estimated employing Saha's equation 1.1:

$$\frac{n_e n_i}{n_a} = \frac{2(2\pi m_e k_b T)^{3/2}}{h^3} \frac{Q_i(T)}{Q_a(T)} e^{-\frac{E_i}{k_b T}},$$

where a two temperature model was assumed. The temperature T was assumed to be the electron temperature T_e [66] and the ratio of the partition function $\frac{Q_i(T)}{Q_a(T)}$ was assumed to be 1 [45]. The overall ion density is given by:

$$n_{i,overall} = \sum_k n_{i,k} = \sum_k \frac{n_{a,k}}{n_e} \frac{2(2\pi m_e k_b T_e)^{3/2}}{h^3} e^{-\frac{E_{i,k}}{k_b T_e}} \tag{4.20}$$

and $n_i \cdot n_e$ becomes:

$$n_i \cdot n_e = const\, T_e^{3/2} \sum_k n_{a,k} e^{-\frac{E_{i,k}}{k_b T_e}} \tag{4.21}$$

with:

$$const = \frac{2(2\pi m_e k_b)^{3/2}}{h^3}. \tag{4.22}$$

Table 4.3: Summary of the determined fundamental plasma parameters of the APS plasma, applied methods, and used transitions.

plasma parameter	applied methods	used transitions
rotational or gas temperature $T_g \approx T_{rot}$	line intensity ratios: comparison of simulated and measured spectra	$A^2\Sigma^+, \nu' = 0$ $\rightarrow X^2\Pi, \nu'' = 0$ of the OH radical
excitation or electron temperature $T_e \approx T_{ex}$	line intensity ratios: Boltzmann plot	atomic oxygen lines
neutral particle density n_a	Dalton's law	
electron density n_e	Saha equation	

Due to the quasi neutrality $n_e = n_i$ the electron density n_e can be calculated by:

$$n_e = \sqrt{const\, T_e^{3/2} \sum_k n_{a,k} e^{-\frac{E_{i,k}}{k_b T_e}}}. \tag{4.23}$$

To summarise, the optical emission spectrum of dry air plasma is dominated by NO-bands in the UV range. Two atomic oxygen lines at 777 nm and 844 nm can be observed in the IR range. These two atomic oxygen transitions were used to obtain the electron temperature T_e. When the air is moistened, additionally to the NO-bands, OH-bands appear in the UV range of the spectrum. This OH-band between 306..310 nm was used to determine the gas temperature T_g. Unfortunately, the neutral particle density n_a and the electron density n_e could not be determined directly by means of optical emission spectroscopy for an air plasma. The neutral particle density n_a was therefore calculated employing Dalton's law 4.19. The Saha equation 1.1 was used to obtain an estimate of the electron density n_e. Thus the four plasma parameters gas and electron temperatures T_g and T_e as well as their densities n_a and n_e were determined. Table 4.3 summarises the methods, which were applied. In the following section 4.2.3 the experimental results from the measurements of the neutral particle and electron temperatures as well as the determination of their densities are presented.

4.2.3 Experimental Results and Interpretations

In the previous section 4.2.2 optical emission spectra of dry and humid air plasmas were shown. Based on these spectra the spectroscopic methods which can be applied to obtain the plasma parameters, the neutral gas and electron temperatures and their densities, were presented. In this section the experimental results of the measurements of the temperatures by means of optical emis-

4.2. CHARACTERISATION OF THE APS PLASMA

sion spectroscopy of dry and humid air plasmas as well as the calculations of the neutral particle and electron densities are presented.

Since the abatement of waste gases is studied for different gas flows and microwave powers the plasma of the APS is characterised in dependence of the supplied microwave power and air flow. A first impression of how the APS plasma behaves in dependence of the microwave power and gas flow can already be studied when the APS is observed without spectroscopic methods. Fig. 4.8 shows photos of the APS for three different gas flows, 10 sl/min, 30 sl/min, and 70 sl/min and three different microwave powers, 1 kW, 2 kW, and 3 kW. For a microwave power of 3 kW only air flows of 30 sl/min and 70 sl/min could be performed since at 10 sl/min the plasma extends to far in radial direction, touching and thereby destroying the quartz tube. The resonator with the slit at its front is located in the lower part of each photo. The quartz tube with a length of 564 mm extends above the resonator and confines the plasma. The air flow was supplied via the tangential gas lead. Therefore, the plasma flame shows a slight torsion for high gas flows of 30 sl/min and 70 sl/min for a microwave power of 1 kW and for an air flow of 70 sl/min and microwave powers of 2 kW and 3 kW. Furthermore, the photos clearly show that the length and the radius of the plasma and therefore the volume of the plasma decrease when the gas flow is increased. On the other hand the plasma radius and length increase when the microwave power is increased.

Further characterisations of the plasma, like the determination of the gas and electron temperature and their densities, were performed by means of optical emission spectroscopy and is described in the following. The gas temperature T_g was measured by using the $A^2\Sigma^+ - X^2\Pi_\gamma$-transition of the free OH-radical. The electron temperature T_e was estimated by the excitation temperature T_{ex}, which was measured by means of a Boltzmann plot from two atomic oxygen lines, as explained in section 4.2.2. All the measurements of the two temperatures are averaged over the line of sight. Thus the temperatures in the centre of the plasma should be about 100 K..200 K higher. The neutral particle density n_a was determined using Dalton's law while the electron density n_e was estimated under the assumption, that the plasma is in partial local thermodynamic equilibrium and by using Saha's equation, as also described in section 4.2.2.

Since the photos in Fig. 4.8 show that the most extended plasma can be observed for a microwave power of 3 kW and a gas flow of 30 sl/min, the radial and axial profiles of the gas rotational temperature T_{rot}, the excitation temperature T_{ex}, and electron density n_e are discussed in the following for this parameter set.

At first gas rotational temperature T_{rot} profiles spatially resolved in axial and radial direction are described.
As already explained in section 4.1.1 the resonator is equipped with a slit at its front so that the plasma can be observed inside the resonator. The resonator has a height of 48 mm and is terminated by a water-cooled metal plate which is why, the characterisation of the plasma in axial

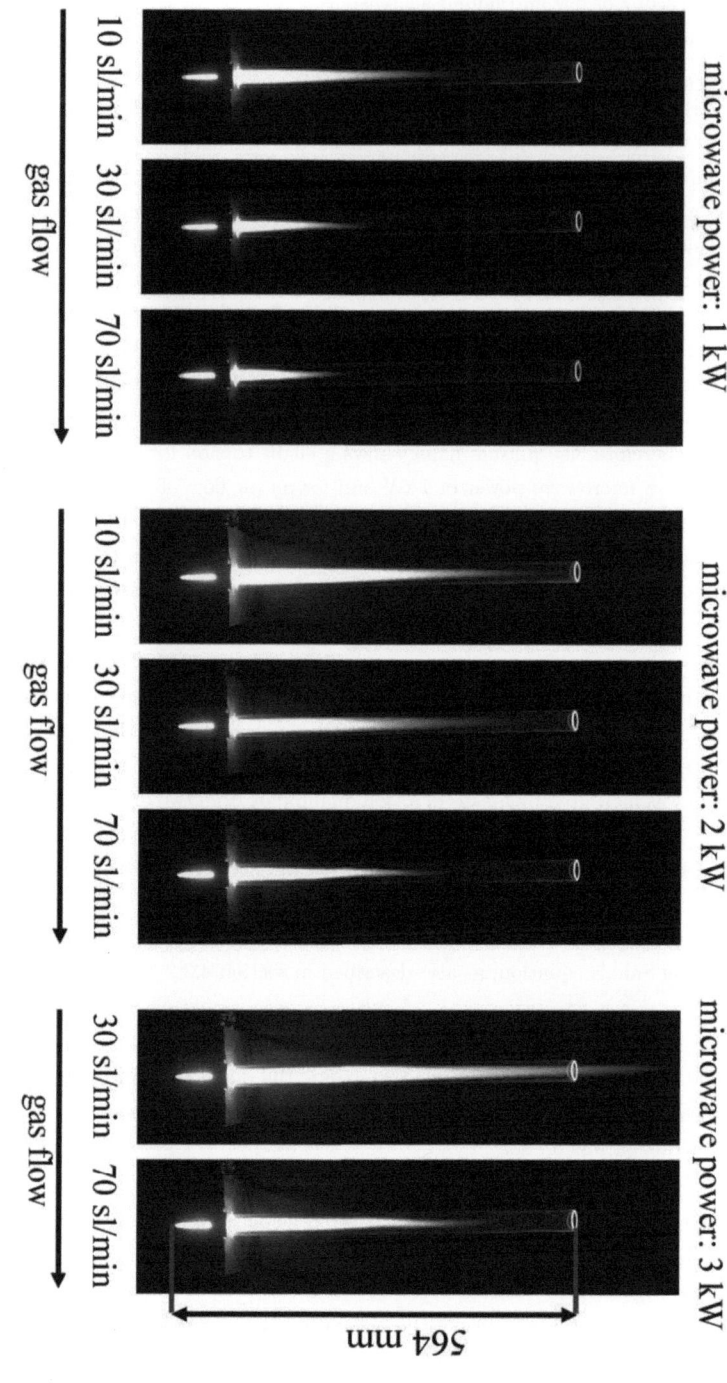

Figure 4.8: Photos of the APS plasma for three different microwave powers: 1 kW, 2 kW, and 3 kW and three different air flows: 10 sl/min, 30 sl/min, and 70 sl/min. It can be seen that the length as well as the radius of the plasma increase with an increase of the supplied microwave power whereas the plasma length and radius decrease with an increasing gas flow.

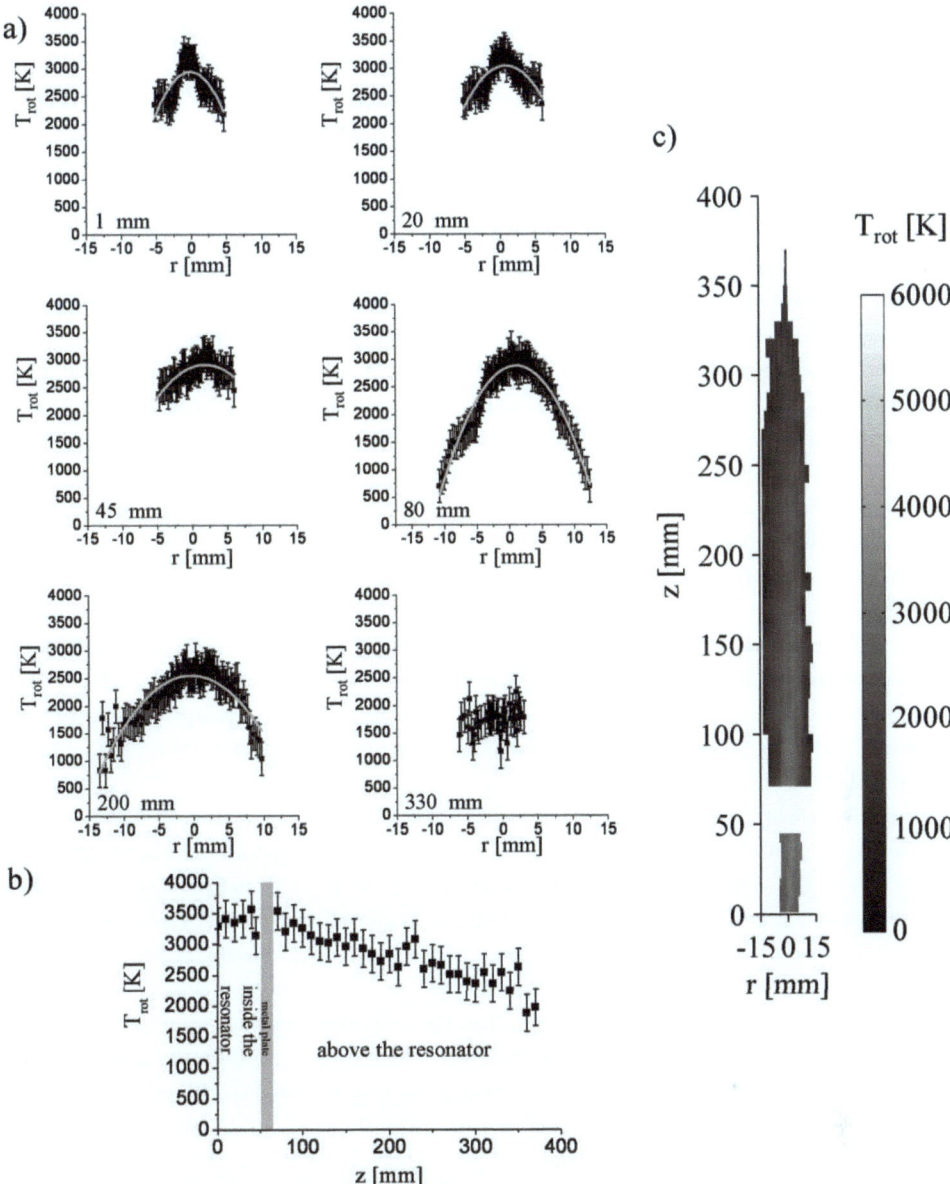

Figure 4.9: Profiles of the gas rotational temperature T_{rot} spatially resolved in axial and radial direction for a microwave power of 3 kW and an air flow of 30 sl/min: a) Radial profiles of T_{rot} at axial positions of: 1 mm, 20 mm, and 45 mm in the resonator, and at 80 mm, 200 mm, and 330 mm above the resonator. The radial profiles were fitted with a parabola. b) An axial profile of T_{rot} at a radial position of $r = 0$ mm. c) A contour plot of T_{rot} spatially resolved in axial and radial direction.

direction is only possible for $z = 1\,\text{mm}..48\,\text{mm}$ and again from $z > 70\,\text{mm}$. The resonator base is located at $z = 0\,\text{mm}$. Since the plasma is confined in a quartz tube with an inner diameter of $d = 26\,\text{mm}$ the extent of the plasma in radial direction is limited to this range and therefore, the radial profiles are shown for $r - 15\,\text{mm}..15\,\text{mm}$. Although the first lens was placed just behind the slit, inside the resonator only the radial region from $r = -5\,\text{mm}$ to $r = 5\,\text{mm}$ was accessible in the UV range. Since the Acton spectrometer is equipped with an ICCD camera, each radial profile could be measured in one step.

Selected radial profiles of the gas rotational temperature T_{rot} are shown in Fig. 4.9 a). The profiles at 1 mm, 20 mm, and 45 mm were measured in the resonator while the profiles at 80 mm, 200 mm, and 330 mm were measured above the resonator. The radial profiles were fitted with a parabola. The asymmetry, which can be observed in all profiles, can be explained by the fact that the gas is supplied via the tangential gas lead, which leads to a slight swirl of the gas (compare Fig. 4.8). The radial profile at a height of $z = 1\,\text{mm}$ shows a maximum of T_{rot} of 3290 K around $r \approx 0$. Outwards T_{rot} first decreases sharply and then the profile flattens. A similar radial profile was measured at a height of $z = 20\,\text{mm}$ but with a slightly higher temperature of $T_{rot} = 3350\,\text{K}$. The profile measured at a height of $z = 45\,\text{mm}$ shows a different behaviour: the maximum in the centre is no longer that distinct.

Above the resonator the whole radial region is accessible. The radial temperature profile at an axial position of $z = 80\,\text{mm}$ has its maximum of $T_{rot} = 3200\,\text{K}$ in the centre, T_{rot} decreases outwards, and can be measured in axial direction in the range of $r = -12..12.5\,\text{mm}$. A similar profile is observed at $z = 200\,\text{mm}$ but with a decreased maximum temperature of $T_{rot} = 2840\,\text{K}$ in the centre. At an axial position of $z = 330\,\text{mm}$ the profile has changed. A temperature can only be measured in the radial region of $r = -6..3\,\text{mm}$, which is almost constant at about $T_{rot} = 1800\,\text{K}$. A parabola could no longer be fitted here.

The distribution of the gas rotational temperature in axial direction at $r = 0\,\text{mm}$ is shown in Fig. 4.9 b). It can be seen that the temperature first grows inside the resonator from $T_{rot} = 3290\,\text{mm}$ at $z = 1\,\text{mm}$ to $T_{rot} = 3560\,\text{K}$ at $z = 40\,\text{mm}$ and than drops to $T_{rot} = 3140\,\text{K}$ at $z = 45\,\text{mm}$. Above the resonator $T_{rot} = 3530\,\text{K}$ can be measured at $z = 71\,\text{mm}$. Then T_{rot} decreases slowly upwards and can be measured until $z = 370\,\text{mm}$ where T_{rot} reaches 1970 K. Since a higher T_{rot} was measured just above the metal plate of the resonator than at $z = 45\,\text{mm}$, the measurement at 45 mm seems to be affected by stronger errors more than the other measurements.

Measurements of the gas rotational temperature were not only performed for the profiles shown in Fig. 4.9 a) and b) but also for radial profiles along the axial direction in intervals of 10 mm. All these measurements are presented as a contour plot, which shows T_{rot} of the plasma flame spatially resolved in axial and radial direction, in Fig. 4.9 c).

Besides the gas rotational temperature T_{rot} the excitation temperature T_{ex} was measured. Since the necessary atomic oxygen lines are located in the IR range and since the refractive power of the lens is dependent of the wavelength, the optical setup had to be adjusted for these measurements. This led to the pleasant circumstance, that the whole radial region from $r = -13\,\text{mm}..13\,\text{mm}$ was accessible. In axial direction again the region between $z = 1..48\,\text{mm}$ and $z > 70\,\text{mm}$ could be measured.

In Fig. 4.10 a) selected radial profiles of the excitation temperature are shown. The profiles at $z = 1\,\text{mm}$, $z = 20\,\text{mm}$, and at $z = 45\,\text{mm}$ were recorded inside the resonator while the profiles at $z = 71\,\text{mm}$, $z = 90\,\text{mm}$ and $z = 110\,\text{mm}$ were measured above the resonator. Like the T_{rot} profiles the radial profiles of T_{ex} were also fitted with a parabola.

At an axial position of $z = 1\,\text{mm}$ T_{ex} can only be measured in the radial range of $r = -1\,\text{mm}..1\,\text{mm}$. At $r = 0\,\text{mm}$ T_{ex} reaches a maximum of $T_{ex} = 3850\,\text{K}$. At $z = 20\,\text{mm}$ the radial range where T_{ex} could be measured has increased to $r = -12..9\,\text{mm}$. Furthermore, the maximum temperature in the centre has also increased to $T_{ex} = 5450\,\text{K}$. At a height of $z = 45\,\text{mm}$ the maximum temperature of T_{ex} has dropped to $T_{ex} = 4820\,\text{K}$, but the radial range stays nearly the same. Above the resonator, at $z = 71\,\text{mm}$, T_{ex} further drops to $T_{ex} = 4600\,\text{K}$, whereas the radial range again stays the same. At an axial position of $z = 90\,\text{mm}$ the radial range where T_{ex} could be measured has decreased to $r = -4\,\text{mm}..4\,\text{mm}$ and the maximum temperature has also decreased to $T_{ex} = 4030\,\text{K}$. At $z = 110\,\text{mm}$ T_{ex} could only be measured around $r = 0\,\text{mm}$ with temperatures of $T_{ex} = 1760\,\text{K}..1930\,\text{K}$.

In Fig. 4.10 b) T_{ex} in dependence of the axial position z at a radial position of $r = 0\,\text{mm}$ is shown. It can be seen that the excitation temperature T_{ex} inside the resonator first increases from $T_{ex} = 3850\,\text{K}$ at $z = 1\,\text{mm}$ to $T_{ex} = 5780\,\text{K}$ at $z = 30\,\text{mm}$. Then the temperature drops to $T_{ex} = 4820\,\text{K}$ at $z = 45\,\text{mm}$. Above the resonator T_{ex} further decreases from $4600\,\text{K}$ at $z = 71\,\text{mm}$ to $T_{ex} = 1760\,\text{K}$ at $z = 110\,\text{mm}$. Further above the resonator no excitation temperature T_{ex} could be measured.

Beside the radial and axial profiles shown in Fig. 4.10 a) and b), more radial profiles have been measured and are presented in the contour plot shown in Fig. 4.10 c).

The measurement of the gas rotational temperature T_{rot} and excitation temperature T_{ex} showed that, for a microwave power of 3 kW and a gas flow of 30 sl/min, the axial range where T_{rot} could be measured ranges from $z = 1\,\text{mm}$ to $z = 370\,\text{mm}$ and is much larger compared to the range of $z = 1\,\text{mm}..110\,\text{mm}$ where T_{ex} could be determined. Furthermore, these measurements revealed that T_{ex} is about 2000 K higher than T_{rot} and therefore, the plasma is presumably in partial local thermodynamic equilibrium.

The determination of the excitation temperature T_{ex} in the air plasma is based on the evaluation of only two atomic oxygen lines. To verify if the evaluation of these two atomic lines leads to a correct Boltzmann plot, oxygen plasmas, which exhibit more atomic oxygen lines, were studied. Three Boltzmann plots, which contain all atomic oxygen lines, that are observed in the oxygen plasmas, are shown in Fig. 4.11. The measurements were performed for a gas flow of 30 sl/min oxygen and at microwave powers of 1 kW, 2 kW, and 3 kW at $z = 30\,\text{mm}$ and $r = 0\,\text{mm}$. The Boltzmann plots show that the two atomic lines at 777 nm and 844 nm are perfectly located on the straight line and therefore these two atomic oxygen lines can be used to determine the excitation

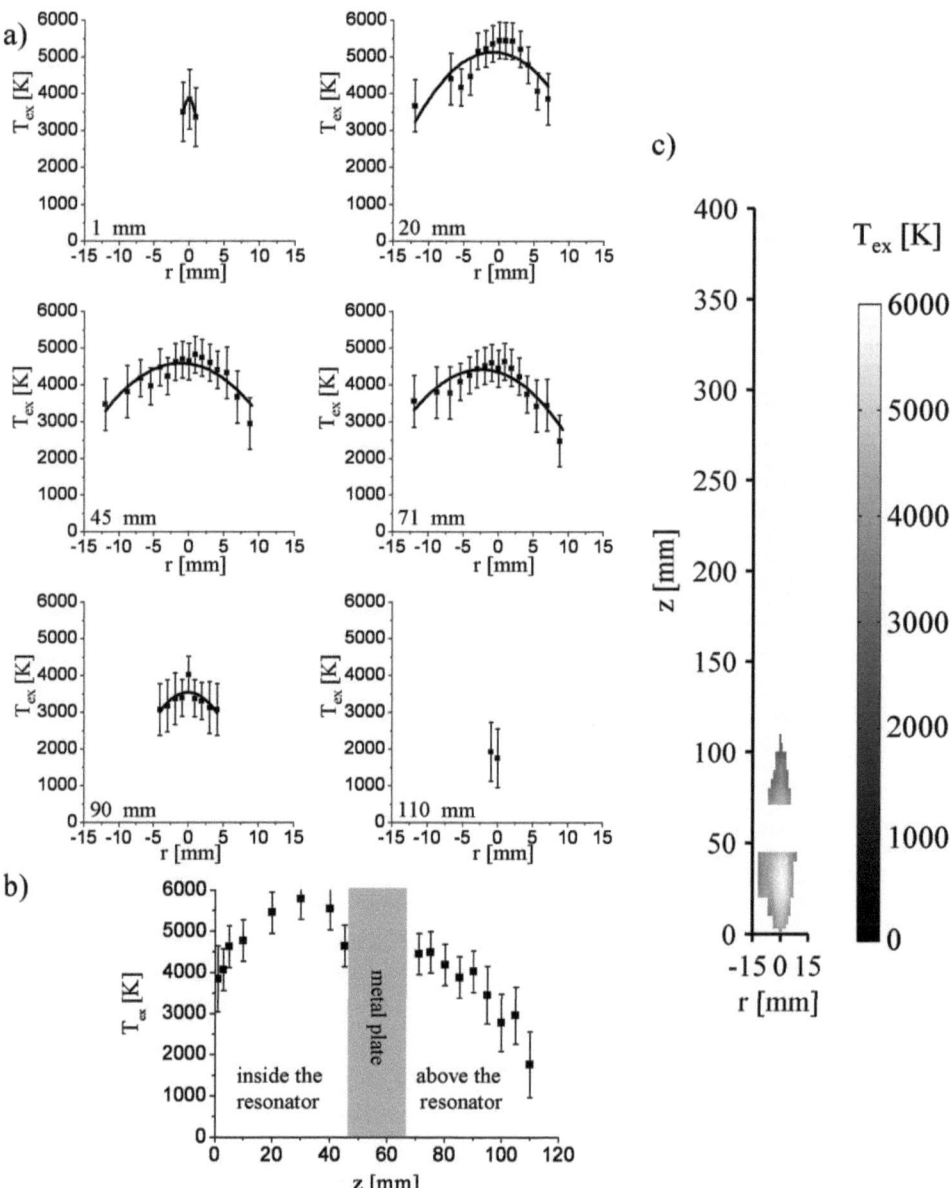

Figure 4.10: Profiles of the excitation temperature T_{ex} spatially resolved in axial and radial direction for a microwave power of 3 kW and an air flow of 30 sl/min: a) Radial profiles of T_{ex} at axial positions of: 1 mm, 20 mm, and 45 mm in the resonator, and at 71 mm, 90 mm, and 110 mm above the resonator. The radial profiles were fitted with a parabola. b) An axial profile of T_{ex} at a radial position of $r = 0$ mm. c) A contour plot of T_{ex} spatially resolved in axial and radial direction.

4.2. CHARACTERISATION OF THE APS PLASMA

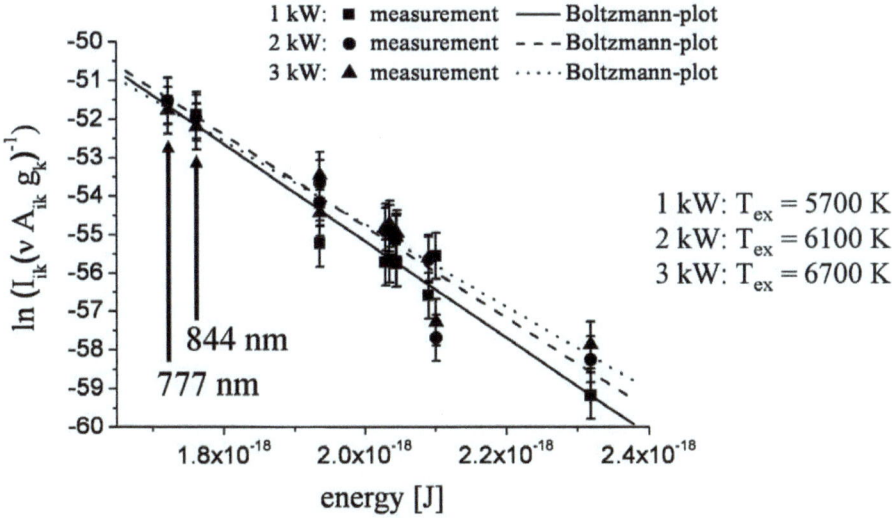

Figure 4.11: Three Boltzmann plots of atomic oxygen lines, which are observed in an oxygen plasma: These measurements were conducted for an oxygen flow of 30 sl/min and at microwave powers of 1 kW, 2 kW and 3 kW and at an axial position of $z = 30$ mm and a radial position of $r = 0$ mm. Since the two atomic lines at 777 nm and at 844 nm are perfectly located on the straight line these two atomic oxygen lines can be used to determine the excitation temperature T_{ex}.

temperature T_{ex}.

The gas and electron temperature $T_g \approx T_{rot}$ and $T_e \approx T_{ex}$ could directly be determined from the measured spectra. By means of equation 4.19 the neutral particle density can now be calculated. The measurements presented in Fig. 4.9 showed that the gas rotational temperature T_{rot} ranges between $T_{rot} = 1880$ K and $T_{rot} = 3560$ K which leads to neutral particle densities of $n_a = 3.9 \cdot 10^{24}$ m^{-3} to $n_a = 2 \cdot 10^{24}$ m^{-3}, respectively.

The electron density n_e can be estimated by Saha's equation as presented in section 4.2.2. For the calculation of the electron density the composition of the air in the plasma was assumed to be $\approx 80\,\%$ nitrogen and $\approx 20\,\%$ oxygen with ionisation energies of 15.58 eV and 12.10 eV, respectively. Furthermore, NO was incorporated since the analyses of the clean gas, which were performed to study the decompostion of waste gases and are presented in section 4.3, revealed that NO is produced in relatively high concentrations of 1300 ppm to 2300 ppm and has a relatively low ionisation energy of 9.26 eV compared to the ionisation energies of nitrogen and oxygen. Therefore, NO was incorporated for the calculation of the electron density with a mid-concentration of $\approx 0.18\,\%$. NO$_2$

and N_2O were neglected since they appear in distinctly lower concentrations and their ionisation energies are higher than the one of NO.

Since the electron density is calculated from the gas rotational and excitation temperature profiles, which are presented in Fig. 4.9 and Fig. 4.10, n_e could only be calculated in regions, where T_{rot} as well as T_{ex} could be measured. Selected radial profiles are shown in Fig. 4.12 a). At an axial position of $z = 1\,\text{mm}$ n_e could only be calculated in a radial range of $r = -1..1\,\text{mm}$. Here a maximum electron density of $n_e = 1.9 \cdot 10^{18}\,\text{m}^{-3}$ is reached. At a height of $z = 20\,\text{mm}$ a maximum electron density of $n_e = 1.7 \cdot 10^2\,\text{m}^{-3}$ is reached and n_e can be calculated from $r = -5\,\text{mm}$ to $r = 5\,\text{mm}$. At $z = 45\,\text{mm}$ the maximum n_e decreases to $n_e = 4.2 \cdot 10^{19}\,\text{m}^{-3}$ and can be calculated from $r = -4..5\,\text{mm}$.

Above the resonator at a height of $z = 71\,\text{mm}$ the electron density can be calculated between $r = -9\,\text{mm}$ and $r = 12\,\text{mm}$ while the maximum n_e around $r = 0\,\text{mm}$ stays at nearly the same value. At an axial position of $z = 90\,\text{mm}$ the radial profile becomes distinctly sharper and the maximum n_e decreases to $n_e = 2.2 \cdot 10^{18}\,\text{m}^{-3}$. At $z = 110\,\text{mm}$ the electron density has dropped further to $n_e = 0.07 \cdot 10^{12}\,\text{m}^{-3}..1.1 \cdot 10^{12}\,\text{m}^{-3}$ and can only be measured around $r = 0\,\text{mm}$.

Fig. 4.12 b) shows the electron density at a radial position of $r = 0\,\text{mm}$ in dependence of the axial position z. It can be seen that n_e increases from $1.9 \cdot 10^{18}\,\text{m}^{-3}$ at $z = 1\,\text{mm}$ to $n_e = 3.2 \cdot 10^{20}\,\text{m}^{-3}$ at $z = 30\,\text{mm}$. Then n_e decreases to $n_e = 4.2 \cdot 10^{19}\,\text{m}^{-3}$ at $z = 45\,\text{mm}$. n_e further decreases above the resonator to $n_e = 1.1 \cdot 10^{12}\,\text{m}^{-3}$ at $z = 110\,\text{mm}$.

n_e was also calculated for all measured T_{rot} and T_{ex} profiles and is presented as a contour plot in Fig. 4.12 c).

The optical emission spectroscopic studies revealed that for an air flow of $30\,\text{sl/min}$ and a microwave power of $3\,\text{kW}$ a maximum gas rotational temperature of $T_{rot} = 3560\,\text{K}$ and a maximum excitation temperature of $T_{ex} = 5780\,\text{K}$ are measured. From these temperatures the neutral particle and electron density can be calculated and the resulting neutral particle density is located in the range of $n_a \approx 10^{24}\,\text{m}^{-3}$ while the electron density is in the range of $n_e \approx 10^{20}\,\text{m}^{-3}$ and therefore is higher than the cutoff density of $n_c = 7.4 \cdot 10^{16}\,\text{m}^{-3}$. Furthermore, the degree of ionisation $\chi = \frac{n_e}{n_a}$ can be calculated and reaches a maximum value of about $\chi \approx 10^{-4}$.

At the beginning of this section 4.2.3 a first impression of how the plasma behaves in dependence of the microwave power and gas flow is presented in Fig. 4.8. For an air flow of $30\,\text{sl/min}$ and a microwave power of $3\,\text{kW}$ a detailed characterisation of the gas and electron temperatures and their densities in the plasma is already presented above. For the same parameter sets as for the plasmas shown in Fig. 4.8, measurements of the gas rotational and excitation temperature and calculations of the electron densities spatially resolved in axial and radial direction were performed which are presented below.

Fig. 4.13 shows contour plots of T_{rot} for microwave powers of $1\,\text{kW}$, $2\,\text{kW}$, and $3\,\text{kW}$ and gas flows of $10\,\text{sl/min}$, $30\,\text{sl/min}$, and $70\,\text{sl/min}$ spatially resolved in axial and radial direction. The

4.2. CHARACTERISATION OF THE APS PLASMA

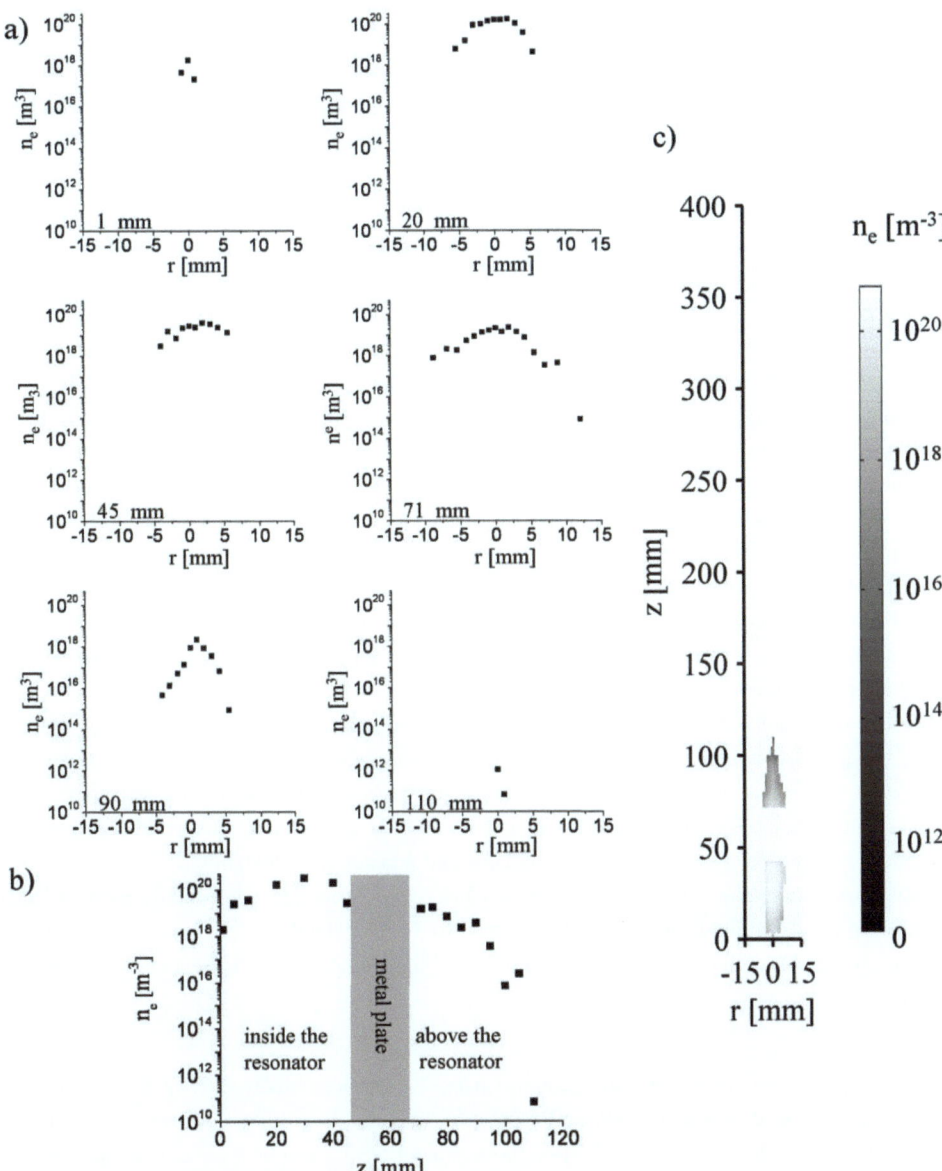

Figure 4.12: Profiles of the electron density n_e spatially resolved in axial and radial direction for a microwave power of 3 kW and an air flow of 30 sl/min: a) Radial profiles of n_e at an axial positions of: 1 mm, 20 mm, and 45 mm in the resonator and at 71 mm, 90 mm, and 110 mm above the resonator. b) An axial profile of n_e at a radial position of $r = 0$ mm. c) A contour plot of n_e spatially resolved in axial and radial direction.

rotational temperature T_{rot} can already be measured at an axial position of $z = 1\,\text{mm}$ for all analysed microwave powers and gas flows. However, the axial extent above the resonator as well as the radial region where T_{rot} can be measured differs in dependence of the microwave power and gas flow. A reasonable statement about the radial extent where T_{rot} can be measured in dependence of the microwave power and gas flow can only be made above the resonator since the radial region where T_{rot} is accessible inside the resonator is limited by the observation slit. Therefore, the radial extent in dependence of the gas flow and microwave power is regarded above the resonator at an axial position of $z = 71\,\text{mm}$. Furthermore, the radial region where T_{rot} can be determined has still the largest extent just above the resonator for all studied microwave powers and gas flows.

At a low microwave power of $1\,\text{kW}$ and a low gas flow of $10\,\text{sl/min}$, the axial region where T_{rot} can be measured ranges up to $z = 210\,\text{mm}$. The radial region at $z = 71\,\text{mm}$ occupies an extent of $R = 10.5\,\text{mm}$. When the gas flow is increased to $70\,\text{sl/min}$ for a microwave power of $1\,\text{kW}$ the axial region where T_{rot} can be determined decreases to $z = 150\,\text{mm}$ and the radial region decreases to $R = 7.6\,\text{mm}$. Thus the axial and radial extent decrease when the gas flow is increased.

On the other hand the axial and radial region increases when the microwave power is increased. This can be seen when the microwave power is increased from $1\,\text{kW}$ to $3\,\text{kW}$ at a gas flow of $30\,\text{sl/min}$. The axial region where T_{rot} can be measured increases from $z = 180\,\text{mm}$ to $z = 370\,\text{mm}$. The radial extent also increases from $R = 8.6\,\text{mm}$ to $R = 11.6\,\text{mm}$. Thus the axial and radial extent increase when the microwave power is increased and decrease when the gas flow is increased. The same behaviour of the plasma in dependence of the supplied microwave power and gas flow could already be observed in the photos, which are shown in Fig. 4.8.

Since the axial and radial extent increase with an increase of the microwave power and decrease when the gas flow is increased, the volume of the plasma also increases with an increase of the microwave power and decreases with a decrease of the gas flow. Thus the minimal plasma volume is observed for a microwave power of $1\,\text{kW}$ and an air flow of $70\,\text{sl/min}$ while the largest plasma volume is reached for a microwave power of $3\,\text{kW}$ and a gas flow of $30\,\text{sl/min}$.

Furthermore, maximum values of the gas rotational temperature T_{rot} are located around a radial position of $r = 0\,\text{mm}$ and range from $T_{rot} = 3000$ to $T_{rot} = 3500\,\text{K}$. Later on, the dependence of T_{rot} on the gas flow and microwave power is discussed and compared to the excitation temperature T_{ex} in detail.

The contour plots of the excitation temperature T_{ex} spatially resolved in axial and radial direction for microwave powers of $1\,\text{kW}$ and $2\,\text{kW}$, and $3\,\text{kW}$ and for air flows of $10\,\text{sl/min}$, $30\,\text{sl/min}$, and $70\,\text{sl/min}$ are shown in Fig. 4.14. Here the region where T_{ex} can be measured is much smaller compared to the region where T_{rot} can be measured. In radial direction T_{ex} cannot be measured from an axial position of $z = 0\,\text{mm}$ for all analysed microwave powers and gas flows. For a low microwave power of $1\,\text{kW}$ T_{ex} can only be measured above $z = 10\,\text{mm}$ up to $z = 71\,\text{mm}$ for gas flows of $30\,\text{sl/min}$ and $10\,\text{sl/min}$ and up to $z = 75\,\text{mm}$ for a gas flow of $70\,\text{sl/min}$. The radial extent is again regarded at an axial position of $z = 71\,\text{mm}$. At a low gas flow of $10\,\text{sl/min}$ the radial region where T_{ex} could be measured is limited to $R = 3\,\text{mm}$. When the gas flow is increased to

4.2. CHARACTERISATION OF THE APS PLASMA

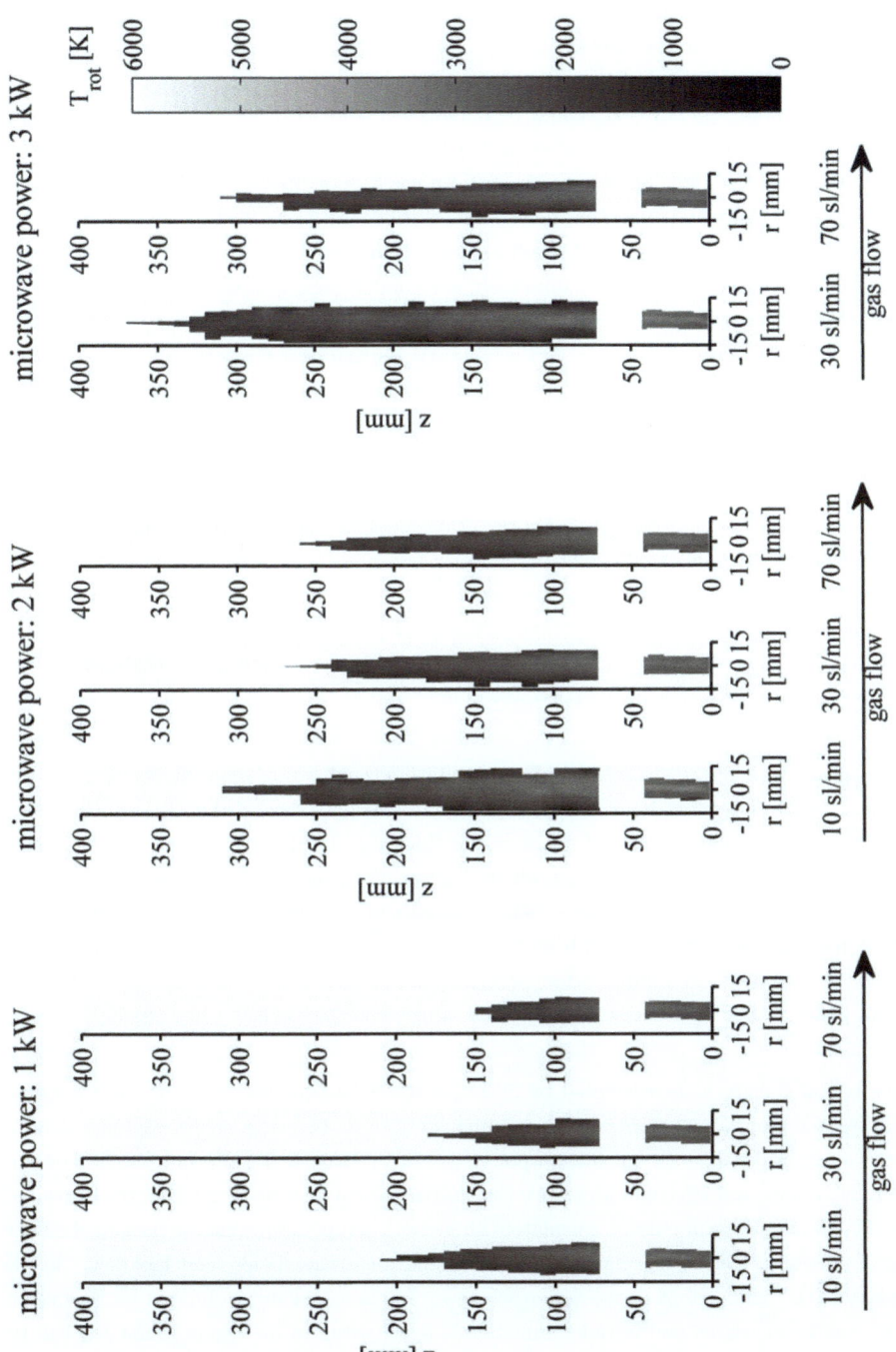

Figure 4.13: Contour plots of the gas rotational temperature T_{rot} for different microwave powers and air flows spatially resolved in axial and radial direction. The measurements were performed for gas flows of 10 sl/min, 30 sl/min, and 70 sl/min air and microwave powers of 1 kW, 2 kW, and 3 kW.

70 sl/min, the radial extent decreases to $R = 2.4$ mm. Thus the radial region where T_{ex} can be measured decreases with an increase of the gas flow but the axial extent stays almost the same and therefore, is independent of the gas flow.

For a microwave power of 2 kW, T_{ex} could be determined in axial direction from $z = 3$ mm for gas flows of 10 sl/min and 30 sl/min and from $z = 10$ mm for a gas flow of 70 sl/min. T_{ex} can be measured above the resonator up to $z = 90$ mm for a gas flow of 10 sl/min and up to $z = 95$ mm for gas flows of 30 sl/min and 70 sl/min. Thus an increase of the power leads to an increase of the axial extent in both axial directions. The radial extent at an axial position of $z = 71$ mm decreases form $R = 10.4$ mm at a gas flow of 10 sl/min to $R = 5.5$ mm at a gas flow of 70 sl/min. This shows that the radial extent also decreases when the gas flow is increased for a microwave power of 2 kW. When the microwave power is increased further to 3 kW, T_{ex} can be determined from $z = 1$ mm up to $z = 110$ mm. The axial extent has again increased in both axial directions. The radial extent at $z = 71$ mm again decreases from $R = 10.4$ mm at a gas flow of 30 sl/min to $R = 8.8$ mm at a gas flow of 70 sl/min.

Thus the contour plots show that the axial and radial extent where T_{ex} can be measured increase when the microwave power is increased. On the other hand the radial extent decreases when the gas flow is increased but the axial extent stays nearly the same.

Furthermore, the maximum values of T_{ex} were again measured around a radial position of $r = 0$ mm and at an axial position of $z = 30$ mm. These values range from $T_{ex} = 5200$ K to $T_{ex} = 5780$ K and are about 2000 K higher than the maximum values of T_{rot}. A detailed discussion of the excitation temperature T_{ex} in dependence of the microwave power and gas flow is presented later on.

The measurements of the gas rotational and excitation temperature spatially resolved in axial and radial direction for different microwave powers and gas flows showed that T_{rot} can be determined in a distinctly larger axial and radial region than T_{ex}. Furthermore, these measurements showed, that the axial and radial region where T_{rot} can be determined increases when the microwave power is increased and decreases with an increase of the gas flow. The axial region where T_{ex} can be measured also increases with an increase of the microwave power and decreases when the gas flow is increased. The radial extent where T_{ex} can be determined also increases when the microwave power is increased but stays nearly the same when the gas flow is increased.

The electron density n_e is calculated by means of the Saha equation from the gas rotational temperature and excitation temperature as explained above. Fig. 4.15 shows contour plots of calculated electron densities for microwave powers of 1 kW, 2 kW, and 3 kW and for gas flows of 10 sl/min, 30 sl/min, and 70 sl/min air. Since the electron density is calculated from the measured T_{rot} and T_{ex} profiles, n_e can only be calculated where both temperatures can be measured. Thus the region where n_e can be calculated depends on the microwave power and gas flow. Inside the resonator, the axial extent of the region where n_e can be calculated is limited by the region where T_{ex} can be measured and in radial direction it is restricted to the region where T_{rot} can be determined. Above the resonator, the region where T_{ex} can be measured limits the region in radial

4.2. CHARACTERISATION OF THE APS PLASMA

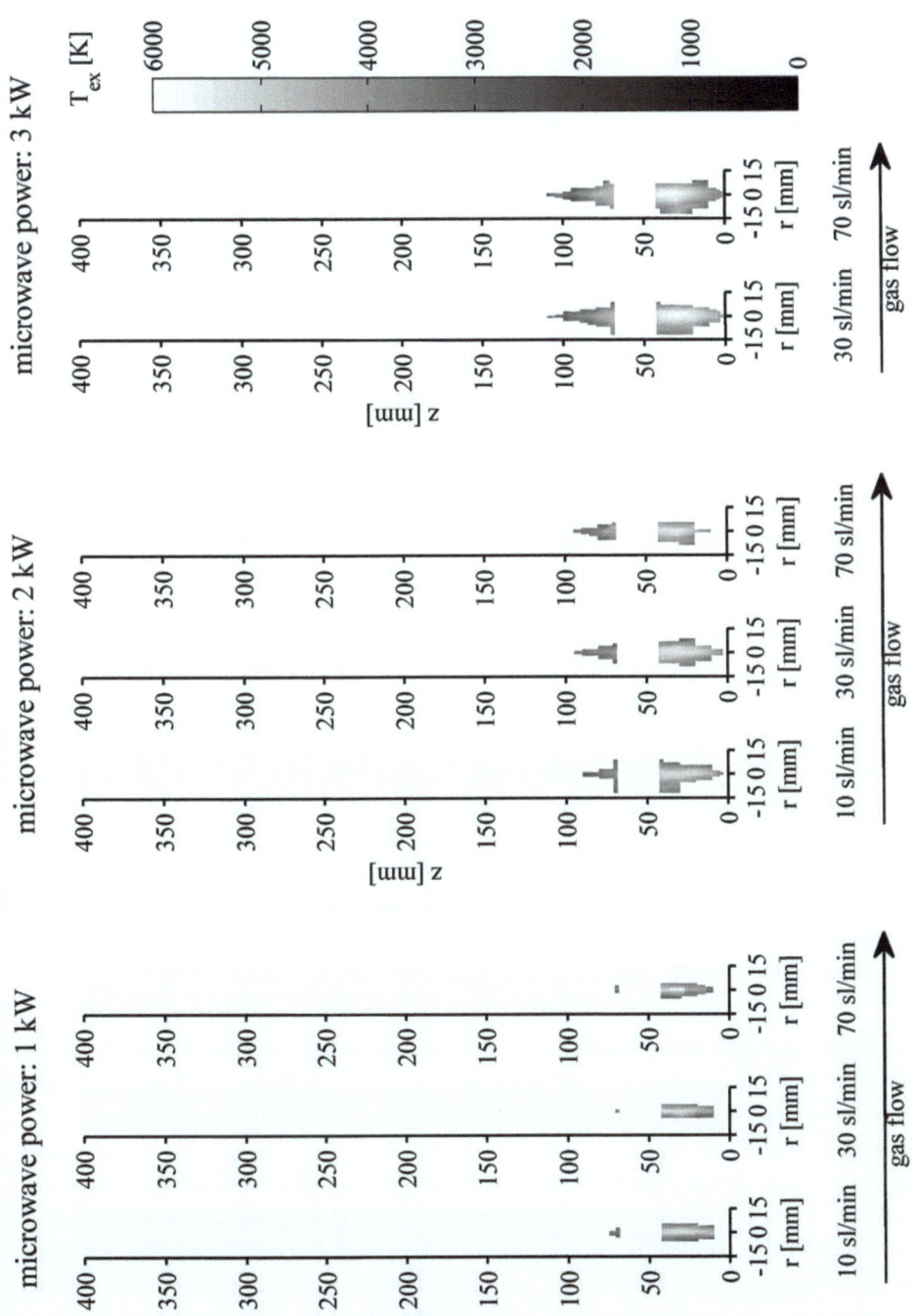

Figure 4.14: Contour plots of the excitation temperature T_{ex} for different microwave powers and air flows spatially resolved in axial and radial direction. The measurements were performed for gas flows of 10 sl/min, 30 sl/min, and 70 sl/min air and microwave powers of 1 kW, 2 kW, and 3 kW.

as well as in axial direction where n_e can be calculated since T_{ex} can be determined in a distinctly smaller region than T_{rot}. Thus the axial and radial extent where n_e can be calculated behaves the same way in dependence of the microwave power and gas flow as the axial and radial extent where T_{ex} was determined.

Since the electron density is calculated by means of the Saha equation it is not surprising that the maximum values of n_e are again reached at an axial position of $z = 30$ mm and around a radial position of $r = 0$ mm. The maximum values of n_e range from $n_e = 1 \cdot 10^{20}$ m^{-3} to $n_e = 3 \cdot 10^{20}$ m^{-3}. The highest electron densities are reached at a gas flow of 30 sl/min and a microwave power of 3 kW.

The measurements of the excitation and gas rotational temperature spatially resolved in axial and radial direction and for three different microwave powers and gas flows showed that maximum gas rotational temperatures of $T_{rot} = 3000$ K..3500 K and about 2000 K higher maximal excitations temperatures of $T_{ex} = 5200$ K..5780K are observed. The gas rotational temperature gives an estimation of the gas temperature T_g and the excitation temperature an lower estimation of the electron temperature T_e. Thus, according to equation 4.19, a neutral particle density of about $n_a = 2 \cdot 10^{24}$ m^{-3}..$4 \cdot 10^{24}$ m^{-3} can be calculated from T_{rot}. The electron density which can be calculated from the gas and electron temperature by means of the Saha equation 1.1 reaches maximum values of $n_e \approx 3 \cdot 10^{20}$ m^{-3}.

These measurements also show that the region where T_{rot} can be measured is much larger than that where T_{ex} can be determined. Furthermore the axial and radial extent of the region where T_{rot} can be determined increases when the microwave power is increased and decreases with an increase of the gas flow. Thus the plasma volume increases with an increase of the microwave power and a decrease of the gas flow. This behaviour of the plasma dependence of the microwave power and gas flow could already be observed in the photos in Fig. 4.8 without any spectroscopic methods.

The axial region where T_{ex} can be measured also increases when the microwave power is increased. However, when the gas flow is increased the radial extent where T_{ex} can be measured decreases while the axial extent stays the same.

Since the behaviour of the maximum values of the excitation and gas rotational temperatures in dependence of the microwave power and of the gas flow cannot easily be seen from the contour plots, the dependence of T_{ex} and T_{rot} on the microwave power and gas flow inside the resonator and above the resonator are discussed separately in detail in the following. Two characteristic positions were chosen. The first position was selected inside the resonator and the other just above the resonator. Inside the resonator an axial position of $z = 30$ mm was chosen since T_{ex} had its maximal values at this position for all analysed microwave powers and gas flows. Since the excitation temperature can only be measured just above the resonator for some parameter sets, an axial position of $z = 71$ mm was chosen as second position. Since the highest temperatures were measured in the centre, a radial position of $r = 0$ mm was selected for both positions.

4.2. CHARACTERISATION OF THE APS PLASMA

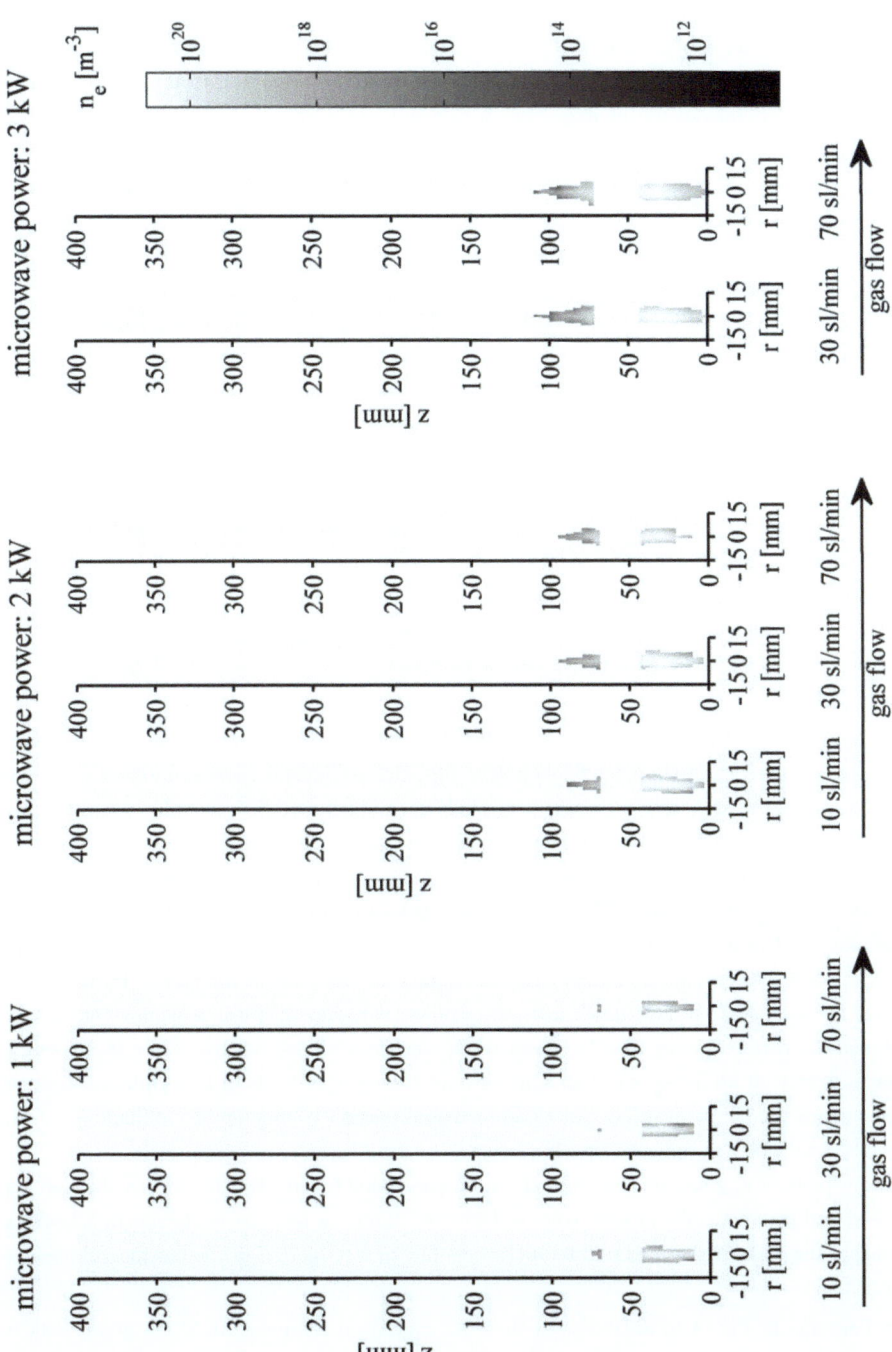

Figure 4.15: Contour plots of the electron density n_e for different microwave powers and air flows spatially resolved in axial and radial direction. The calculations were performed for gas flows of 10 sl/min, 30 sl/min, and 70 sl/min air and microwave powers of 1 kW, 2 kW, and 3 kW.

Fig. 4.16 shows the dependence of the gas rotational temperature T_{rot} and of the excitation temperature T_{ex} on the microwave power and gas flow inside the resonator at an axial position of $z = 30$ mm. The dependency on the microwave power and on the gas flow is exemplarily shown for gas flows of 30 sl/min and 70 sl/min and for microwave powers of 1 kW and 2 kW since for these parameter sets measurements of all microwave powers and gas flows could be conducted. The diagrams in Fig. 4.16 a) and b) show the dependence of T_{rot} and T_{ex} on the supplied microwave power for a gas flow of 30 sl/min and 70 sl/min, respectively. Both diagrams show that, the excitation temperature T_{ex} increases very slightly from $T_{ex} = 5390$ K to $T_{ex} = 5780$ K for a gas flow of 30 sl/min and from $T_{ex} = 5400$ K to $T_{ex} = 5730$ K for a gas flow of 70 sl/min when the microwave power is increased. For a gas flow of 30 sl/min the gas rotational temperature T_{rot} also increases very slightly in dependence of the microwave power from $T_{rot} = 3110$ K to $T_{rot} = 3410$ K while at a gas flow of 70 sl/min T_{rot} is nearly independent of the microwave power and has a value of $T_{rot} = 3020$ K.

The dependence of T_{rot} and T_{ex} on the gas flow is shown in Fig. 4.16 c) and d). These diagrams show that T_{ex} increases very slightly from $T_{ex} = 5190$ K to $T_{ex} = 5400$ K at a microwave power of 1 kW and from $T_{ex} = 5240$ K to $T_{ex} = 5450$ K at a microwave power of 2 kW when the gas flow is increased. The gas rotational temperature T_{rot}, in contrast, decreases very slightly from $T_{rot} = 3320$ K to $T_{rot} = 3020$ K at a microwave power of 1 kW and from $T_{rot} = 3490$ K to $T_{rot} = 2960$ K at a microwave power of 2 kW when the gas flow is increased.

Thus according to the diagrams in Fig. 4.16, the excitation temperature T_{ex} increases very slightly with an increase of the microwave power and with an increase of the gas flow. The gas rotational temperature T_{rot} also increases very slightly or stays at the same value when the microwave power is increased and decreases very slightly when the gas flow is increased.

Nevertheless, the dependence of T_{ex} and T_{rot} on the microwave power as well as on the gas flow is very weak. The contour plots of T_{rot} and T_{ex} in Fig. 4.13 and Fig. 4.14, respectively, showed that the axial and radial extent where these temperatures can be measured increases when the microwave power is increased and therefore, the volume of the plasma increases. Furthermore, qualitative measurements of how much microwave power is radiated by the plasma revealed that the radiated microwave power also increases when the supplied microwave power is increased. These facts show that an increase of the supplied microwave power leads to no significant increase of the maximum temperature values but to an increased plasma volume and to an increase of the microwave power which is radiated by the plasma.

The contour plots of T_{rot} also showed that the axial and radial extent decreases when the gas flow is increased and therefore, the plasma volume decreases. Thus an increase of the gas flow leads to no significant decrease of the maximal temperatures but to a reduction of the plasma volume.

The diagrams in Fig. 4.16 also showed that the excitation temperature T_{ex} is about 2000 K higher than the gas rotational temperature T_{rot}. This shows that the microwave heats the electrons.

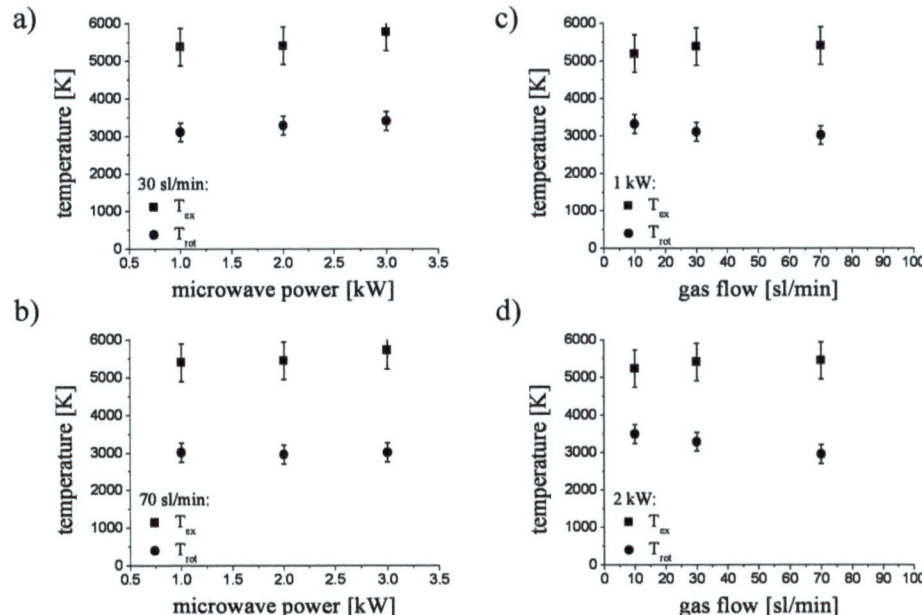

Figure 4.16: Dependence of the gas rotational temperature T_{rot} and of the excitation temperature T_{ex} on the microwave power and on gas flow inside the resonator at an axial position of $z = 30\,\text{mm}$ and a radial position of about $r = 0\,\text{mm}$. Dependence of T_{ex} and T_{rot} on the microwave power at a gas flow a) of $30\,\text{sl/min}$ and b) of $70\,\text{sl/min}$ and c) and d) the dependence of the gas flow on T_{ex} and T_{rot} at a microwave power of $1\,\text{kW}$ and $2\,\text{kW}$, respectively.

This is obvious since only the light electrons are able to follow the quickly oscillating electric field of the microwave. The neutral particles are too heavy to follow the quickly oscillating electric field and therefore are only heated through collisions with the faster electrons.

Information about the energy transfer from the electrons to the heavy particles is provided by the collision frequency. The electron neutral particle collision frequency ν_{en} can be calculated by the following equation [46]:

$$\nu_{en} = n_a v <\sigma_{ea}>, \qquad (4.24)$$

with n_a being the neutral particle density, v the difference velocity of the electrons and the heavy particles, and $<\sigma_{ea}>$ the collisional cross section. The neutral particle density was calculated above and ranges from $n_a = 2 \cdot 10^{24}\,\text{m}^{-3}$ to $n_a = 4 \cdot 10^{24}\,\text{m}^{-3}$. Using the thermal velocity $v_{th} = \sqrt{\frac{2k_b T}{m}}$, with the measured temperatures one gets $v_{the} \approx 10^5\,\frac{\text{m}}{\text{s}}$ for the electrons and $v_{tha} \approx 10^3\,\frac{\text{m}}{\text{s}}$ for the heavy particles. Thus v_{tha} is negligible and the difference velocity is assumed to be v_{the}.

Since air consists of $\approx 80\,\%$ nitrogen, the electron neutral particle frequency can be estimated when the Ramsauer cross section of nitrogen of $<\sigma_{eN_2}> = 6 \cdot 10^{-20}\,\text{m}^{-2}$ is assumed for $<\sigma_{ea}>$ [46]. Then ν_{en} becomes $\nu_{en} \approx 5 \cdot 10^{10}\,\text{s}^{-1}$. However since per collision only a fraction of $\frac{m_e m_a}{(m_e + m_a)^2} \approx 8 \cdot 10^{-5}$

of the enrgy is transfered due to the mass difference, the time for the energy transfer from the electrons to the heavy particles is in the range of microseconds.

The electron electron and neutral particle neutral particle frequencies can also be calculated by equation 4.24 when the electron density, v_{the}, and $<\sigma_{ee}>$ and the neutral particle density, v_{tha}, and $<\sigma_{N_2N_2}>$, respectively, are used. With the measured temperatures and calculated densities ν_{ee} is about $\nu_{ee} \approx 7 \cdot 10^9 \, \text{s}^{-1}$ and $\nu_{N_2N_2} \approx 1 \cdot 10^9 \, \text{s}^{-1}$.

Since ν_{ee} and $\nu_{N_2N_2}$ are in the range of $\approx 10^9 \, \text{s}^{-1}..10^{10} \, \text{s}^{-1}$, the energy transfer between the electrons and between the neutral particles is guaranteed and therefore they obey a Maxwellian velocity distribution. The energy transfer from the electrons to the heavy particles is, in contrast, in the range of microseconds, and therefore the temperature of the electrons should be higher than the temperature of the gas particles which is confirmed by the observations.

Furthermore, the ratio of the electron neutral particle frequency ν_{en} to the microwave frequency ω should be around 1 for a good energy transfer, as explained in chapter 1 section 1.2. With $\nu_{ea} \approx 5 \cdot 10^{10} \, \text{s}^{-1}$ and $\omega = 2\pi \cdot 2.45 \cdot 10^9 \, \text{s}^{-1}$, $\frac{\nu_{ea}}{\omega}$ is about ≈ 3. This means that the plasma is heated by the microwave quite well and therefore the electron temperature is only about 2000 K higher than the gas temperature.

The collisions also enable the microwave to penetrate into the plasma even though maximum electron densities of $n_e = 3 \cdot 10^{20} \, \text{m}^{-3}$ occur which are distinctly higher than the critical density (cutoff density) n_c of $n_c = 7.4 \cdot 10^{16} \, \text{m}^{-3}$. The skin depth δ of the microwave without any collisions is only $\delta = 1.4 \, \text{mm}$, according to equation 1.33, which would result in an inefficient heating of the plasma.

The diagrams in Fig. 4.16 showed that the gas rotational as well as the excitation temperature are almost independent of the supplied microwave power and gas flow. This phenomenon can possibly be explained by the heating mechanism of the microwave. The microwave only heats the light electrons and the heavy particles are heated by collisions with the faster electrons. However, an optimal heating of the electrons by the microwave is only obtained when the electron heavy particle collision frequency ν_{en} is in the range of the microwave frequency ω: $\frac{\nu_{en}}{\omega} \approx 1$.

Assuming that $\frac{\nu_{en}}{\omega}$ is ≈ 1, if then the electron temperature increases, the thermal velocity of the electrons v_{the} would increase, too, which according to equation 4.24 would result in an increased electron heavy particle collision frequency ν_{en}. As a result, the heating of the heavy particles would increase and the heavy particles would reach a higher temperature. A higher temperature of the heavy particles would mean, according to equation 4.19, that the neutral particle density n_a would decrease. However, a decrease of n_a would lead to a decrease of ν_{en} and to a decrease of $\frac{\nu_{en}}{\omega} < 1$ and therefore the heating of the electrons through the microwave would become worse and the electron temperature would decrease to the previous temperature. A decrease of the electron temperature would lead to a decrease of ν_{en} and therefore to a decrease of the gas temperature. If the gas temperature decreases, the neutral particle density n_a would increase. An increase of n_a would result in a increase of ν_{en} and a increase of $\frac{\nu_{en}}{\omega}$ to ≈ 1. This would lead to a better heating of the electrons and therefore the electron temperature would increase to the previous value.

This shows that the heating efficiency of the electrons through the microwave is very strongly

4.2. CHARACTERISATION OF THE APS PLASMA

related to the temperatures. This leads to the fact that the temperatures level out at an optimal value where $\frac{\nu_{en}}{\omega}$ is around 1 and as a result they are almost independent of the supplied microwave power and gas flow. Thus the larger amount of energy which is supplied at a higher microwave power cannot lead to an increase of the temperature and therefore results in a larger volume of the plasma, as already presented above.

Fig. 4.17 shows the dependence of the gas rotational temperature T_{rot} and of the excitation temperature T_{ex} on the microwave power and on the gas flow above the resonator at an axial position of $z = 71$ mm and a radial position of about $r = 0$ mm. The diagrams in Fig. 4.17 a) and b) show the dependence of T_{ex} and T_{rot} on the microwave power for gas flows of 30 sl/min and 70 sl/min, respectively. It can be seen that T_{ex} as well as T_{rot} increase with an increase of the microwave power at that point. T_{ex} increases from $T_{ex} = 3140$ K to $T_{ex} = 4560$ K at a gas flow of 30 sl/min and from $T_{ex} = 3550$ K to $T_{ex} = 4530$ K at a gas flow of 70 sl/min. The gas rotational temperature increases from $T_{rot} = 2810$ K to $T_{rot} = 3070$ K at a gas flow of 30 sl/min and from $T_{rot} = 2620$ K to $T_{rot} = 2950$ K at a gas flow of 70 sl/min.

The dependence of T_{ex} and T_{rot} on the gas flow for a microwave power of 1 kW and 2 kW are shown in Fig. 4.17 c) and d), respectively. For a microwave power of 1 kW T_{ex} increases from $T_{ex} = 3000$ K to $T_{ex} = 3180$ K when the gas flow is increased from 10 sl/min to 70 sl/min. T_{rot} also increases slightly from $T_{rot} = 2560$ K to $T_{rot} = 2390$ K. For a microwave power of 2 kW the same dependence of the two temperature on the gas flow can be observed. T_{ex} increases slightly from $T_{ex} = 3670$ K to $T_{ex} = 4240$ K and T_{rot} from $T_{rot} = 2770$ K to $T_{rot} = 2550$ K.

Above the resonator at an axial position of $z = 71$ mm T_{ex} as well as T_{rot} increase slightly when the microwave power is increased. When the gas flow is increased, T_{ex} also increases slightly but T_{rot} decreases slightly. The increase of T_{ex} and T_{rot} in dependence of the supplied microwave power just above the resonator can be explained thereby that the axial extent where the plasma is heated increases when the microwave power is increased. Thus higher values of T_{rot} and T_{ex} can be measured for higher microwave powers at the same axial position above the resonator. The fact that higher excitation temperatures can be measured at $z = 71$ mm for higher gas flows can be explained thereby that the electrons have a higher axial outreach at higher gas flows and therefore, higher values of T_{ex} are observed at the same axial position above the resonator.

Furthermore, these diagrams show that the gas rotational and excitation temperatures almost have the same values at a low microwave power of 1 kW and a low gas flow of 10 sl/min. This means, that here the plasma has already thermalised. For higher gas flows and higher microwave powers the two temperatures have not yet reached the same value at that position. The excitation temperature is still higher than the gas rotational temperature. Thus in the following it will be studied where the two temperatures have approximately equal values and therefore from which axial position on the plasma has thermalised.

Fig. 4.18 shows the axial position where the excitation temperature and the gas rotational

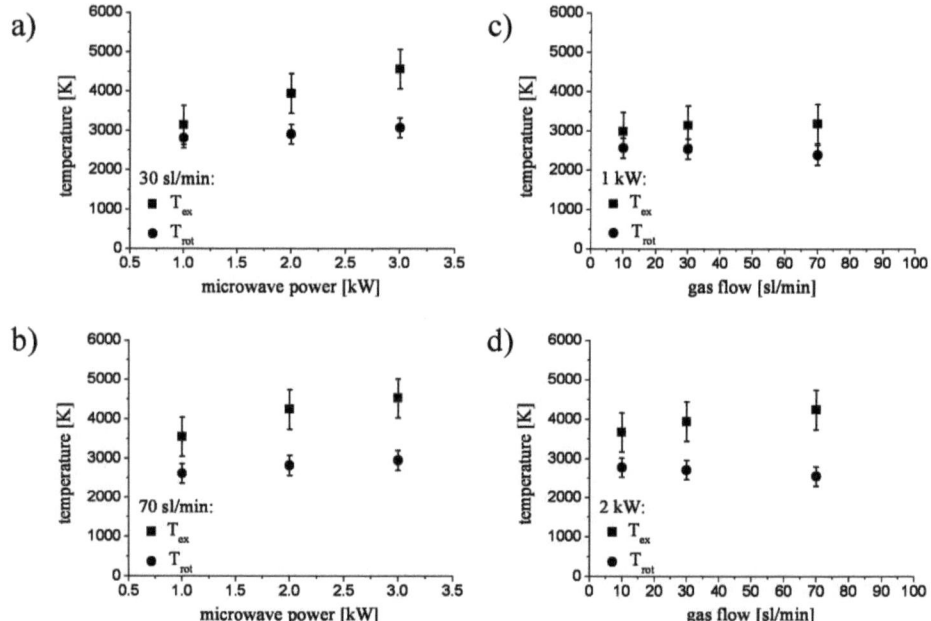

Figure 4.17: Dependence of the gas rotational temperature T_{rot} and of the excitation temperature T_{ex} on the microwave power and on the gas flow above the resonator at an axial position of $z = 71$ mm and a radial position of $r = 0$ mm. Dependence of T_{rot} and T_{ex} on the microwave power at a gas flow of a) 10 sl/min and b) 70 sl/min. Dependence of T_{rot} and T_{ex} on the gas flow at a microwave power of c) 1 kW and d) 2 kW.

temperature have approximately the same value. It can be seen that for a microwave power of 1 kW and gas flows of 10 sl/min and 30 sl/min this point is already reached at an axial position of $z = 71$ mm. For a higher gas flow of 70 sl/min this point is reached at $z = 75$ mm. For a microwave power of 2 kW the point where $T_{ex} \approx T_{rot}$ is reached at $z = 85$ mm for a gas flow of 10 sl/min and at $z = 90$ mm and $z = 95$ mm for higher gas flows of 30 sl/min and of 70 sl/min, respectively. This shows that the point where both temperatures have approximately the same value shifts to higher axial positions for higher microwave powers and for higher gas flows. This also is observed when a microwave power of 3 kW is supplied. Here this point is found at $z = 105$mm for a gas flow of 30 sl/min and at $z = 110$ mm for 70 sl/min.

The fact that the point where T_{ex} and T_{rot} have approximately the same value shifts to higher axial positions for higher gas flows and higher microwave powers can be explained on the one hand thereby that the electrons have a higher axial outreach for higher gas flows and on the other hand that the axial extent where the electrons are heated by the microwave increases with an increase of the microwave power.

4.2. CHARACTERISATION OF THE APS PLASMA

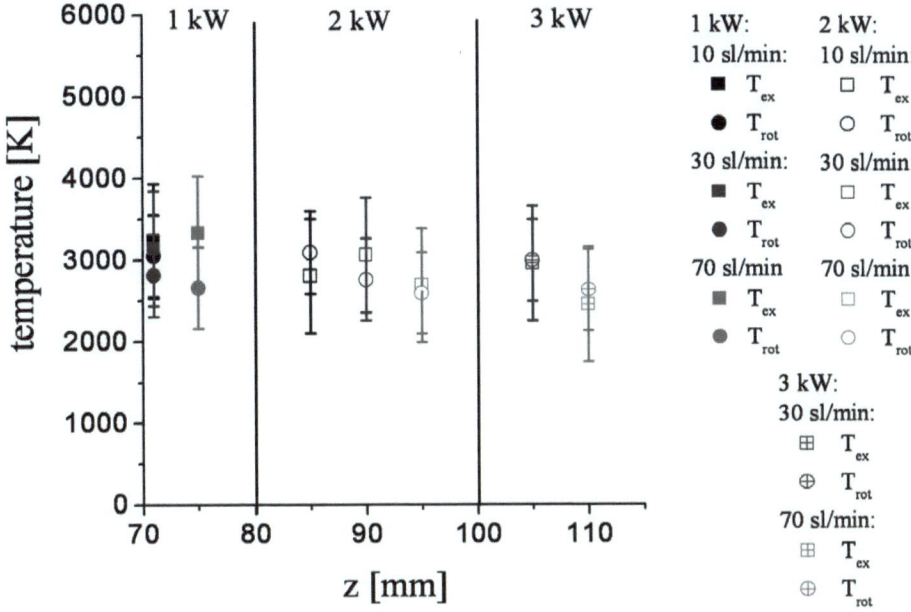

Figure 4.18: Dependence of the axial position where the plasma is thermalising on the microwave power and on the gas flow. The digram shows the axial position z where the excitation temperature T_{ex} has dropped to the gas rotational temperature T_{rot}. The plasma has thermalised at that position. It can be seen that this point drifts to higher values of z when the gas flow or the microwave power is increased.

To summarise, the characterisation of an air plasma without any spectroscopic methods already revealed that the extent of the plasma increases with an increase of the microwave power and decreases when the gas flow is increased, as the photos of the plasma in Fig. 4.8 show.

The overview spectra of an air plasma showed that the spectrum is dominated by NO-bands in the UV. Two oxygen lines at 777 nm and 844 nm were also observed. A humidification of the air flow led to additional OH-bands in the UV range. Thus the gas temperature T_g of the plasma was estimated by the gas rotational temperature T_{rot} which was measured from the $A^2\Sigma^+ - X^2\Pi_\gamma$-transition of the free OH radical. A lower estimation of the electron temperature T_e was provided by the excitation temperature T_{ex} which was measured by means of a Boltzmann plot from the two oxygen lines. The validity of this Boltzmann plot of only two lines was verified by a Boltzmann plot of several oxygen lines which could be observed in an oxygen plasma.

The measurements spatially resolved in axial and radial direction of the gas rotational temperature for three different microwave powers and gas flows showed that the axial and radial extent where T_{rot} can be measured behaves the same way in dependence of the microwave power and gas flows as already observed on the photos. The excitation temperature was also measured for the same parameter sets, which revealed that the axial and radial extent where T_{ex} can be determined increases when the microwave power is increased. When the gas flow is increased, the radial extent also decreases but the axial extent stays the same.

Furthermore, these measurements showed that the maximum values of $T_{rot} = 3000\,\text{K}$ to $3500\,\text{K}$ and of T_{ex} of $T_{ex} = 5200\,\text{K}..5780\,\text{K}$ are reached inside the resonator. The neutral particle density could also be calculated and is about $n_a \approx 10^{24}\,\text{m}^{-3}$. The electron density was estimated by the Saha equation and has maximum values of $n_e \approx 10^{20}\,\text{m}^{-3}$ which results in a maximum ionisation degree of $\chi \approx 10^{-4}$ of the plasma.

Furthermore, the measurements of the temperatures in dependence of the supplied microwave power and the gas flow showed that T_{ex} as well as T_{rot} are only very slightly dependent on these parameters. However, the photos and contour plots showed that the plasma volume increases when the microwave power is increased and decreases with an increase of the gas flow. Thus an increase of the microwave power does not lead to higher temperatures but to a larger plasma volume and on the other hand an increase of the gas flow does not result in a decrease of the temperatures but leads to a reduction of the plasma volume.

Since $T_{ex} \approx T_e$ is about $2000\,\text{K}$ higher than $T_{rot} \approx T_g$, the plasma can be assumed to be in partial local thermodynamic equilibrium. This is obvious since only the light electrons can follow the quickly oscillating electric field of the microwave and therefore only the electrons are heated by the microwave. The heavy particles are heated through collisions with the faster electrons. The collision frequencies are located at $\nu_{ee} \approx 7 \cdot 10^9\,\text{s}^{-1}$, $\nu_{N_2 N_2} \approx 1 \cdot 10^9\,\text{s}^{-1}$, and $\nu_{eN_2} \approx 4 \cdot 10^{10}\,\text{s}^{-1}$. Since the time for the energy transfer between the electrons and between the gas particles is in the range of nanoseconds according to the collision frequencies, the electrons as well as the gas particles obey a Maxwellian velocity distribution. The energy transfer between the electrons and gas particles is reduced by a factor of about $\approx 8 \cdot 10^{-5}$ due to the different masses and thus the time for this energy transfer is in the range of microseconds.

Due to the collisions the microwave can penetrate into the plasma even though the electron density reaches values which are distinctly higher than the critical density (cutoff density) of $n_c = 7.4 \cdot 10^{16}\,\text{m}^{-3}$. Otherwise the skin depth δ of the microwave would only be $\delta = 1.4\,\text{mm}$. Furthermore, the ratio of the electron neutral particle frequency to the frequency of the microwave is about $\frac{\nu_{en}}{\omega} \approx 3$ and therefore near to the value of $\frac{\nu_{en}}{\omega} = 1$ where the heating of the plasma is optimal.

Since the electron heavy particle frequency ν_{en} is very strongly related to the temperatures, the temperatures level out at an optimal value where $\frac{\nu_{en}}{\omega} \approx 1$. This explains why the temperatures are almost independent of the supplied microwave power and gas flow. Thus a larger amount of supplied energy cannot lead to an increase of the maximum temperatures and therefore, results in a larger plasma volume.

Furthermore, it was shown that the point where T_{ex} and T_{rot} have approximately the same value and therefore where the plasma has thermalised, shifts to higher axial positions when the gas flow or the microwave power is increased.

4.3 Decomposition of Waste Gases

In the previous section the characterisation of the APS plasma by means of optical emission spectroscopy was presented. The last part of this chapter covers studies of the suitability of the APS for the decomposition of waste gases, one possible application of the APS. In the following the decomposition of volatile organic compounds (VOC) in air flows on the one hand and of perfluorinated compounds (PFC) in nitrogen flows on the other will be analysed in detail for that purpose.

4.3.1 Abatement of Volatile Organic Compounds

Volatile organic compounds (VOC) emerge during the use and production of, for example lacquer, colouring, and glue. Many of these VOC are harmful and contribute to the world climate change since they are green house gases. Thus the abatement of these gases is an important and urgent task for many industrial branches. In the present work the decomposition of two VOC samples, propane and toluene, are analysed. VOC commonly appear in mixtures with air, which is why the abatement of propane and toluene in air plasmas was studied.

To examine fundamental principles of the decomposition of VOC with the APS, model waste gas mixtures consisting of air and of either propane or toluene were blended. The raw and clean gases were analysed with a FTIR spectrometer, a FID, and a mass spectrometer. These measurements also enabled the calculations of destruction and removal efficiencies (DRE). Additionally samples for a detailed analysis in a gas-phase chromatograph were absorbed on Tenax®. Complementary to the analyses of the raw and clean gases, the plasma was examined by optical emission spectroscopy that provided information about the species, which are present in the plasma. The experimental setup for the decomposition studies is described in detail in section 4.1.2.

In Fig. 4.19 optical emission overview spectra and FTIR spectra of pure air and of air plasma containing propane or toluene are shown. The spectra in Fig. 4.19 a), b), and d) were recorded at an air flow of 12.0 sl/min while the spectra in Fig. 4.19 was recorded at an air flow of 26.0 sl/min. The supplied microwave power was 1 kW for all measurements. In Fig. 4.19 a) and c) propane admixtures of 1.77 mg and 5.90 mg propane and in Fig. 4.19 b) and d) toluene admixtures of 3.70 mg and 12.33 mg were added. These admixtures correspond to a concentration of 900 ppm, 2800 ppm and 405 ppm, 1400 ppm for an air flow of 12.0 sl/min and 26.0 sl/min, respectively.

The optical emission overview spectra were recorded with the Avantes spectrometer at a radial position of $r = 0$ mm and an axial position of $z \approx 30..40$ mm above the resonator base. The overview spectrum of a pure air plasma is dominated in the UV range by the NO$_\beta$- and NO$_\gamma$- bands, which belong to the $B^2\Pi - X^2\Pi$- and $A^2\Sigma^+ - X^2\Pi$-transitions [58]. When an admixture of 1.77 mg propane or 3.70 mg toluene is added, CN-, NH- and OH-bands appear in the spectra additionally to these NO-bands. These bands belong to the $B^2\Sigma - X^2\Sigma$-, the $A^3\Pi - X^3\Sigma^-$,- and the

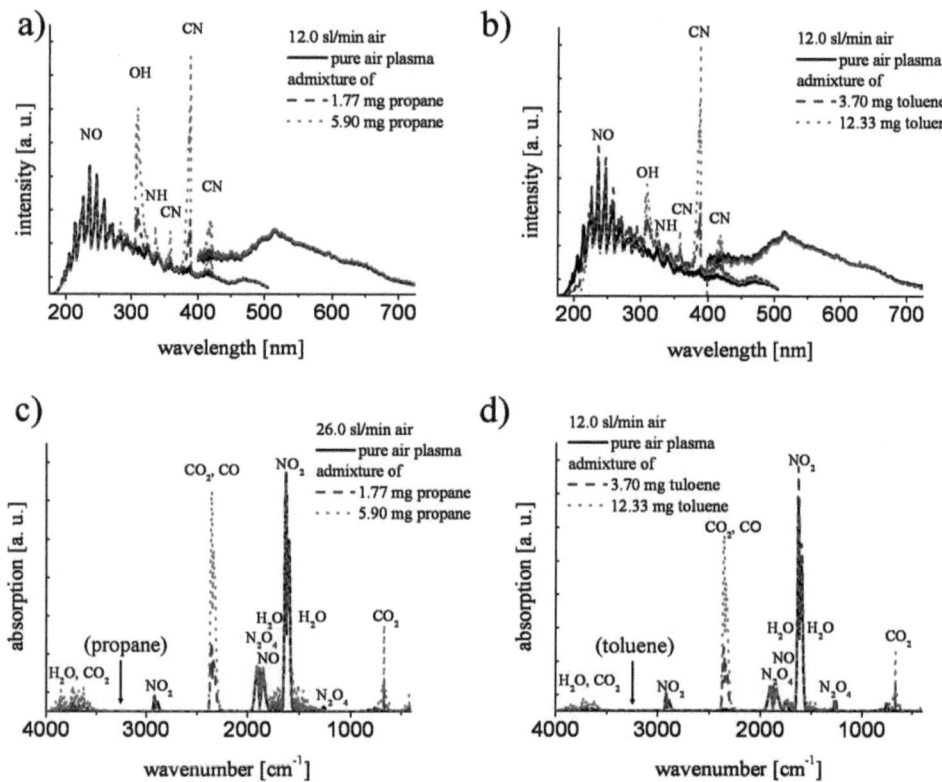

Figure 4.19: Optical emission spectra of a pure air plasma and of an air plasma with an admixture of a) 1.77 mg and 5.90 mg propane and b) 3.70 mg and 12.33 mg toluene for an air flow of 12.0 sl/min and a microwave power of 1 kW. The admixture of propane and toluene corresponds to a concentration of 900 ppm and 2800 ppm. The spectra were recorded with the Avantes spectrometer which is why between 400 nm and 500 nm in each case two spectra are shown. The pure air plasma is dominated by NO-bands in the UV range. Additionally, when propane or toluene is added, CN-, NH-, and OH-bands appear. The NO-bands indicate that nitride oxides are formed while the OH-bands indicate that propane and toluene is decomposed to H_2O.

The diagrams in c) and d) show FTIR spectra of the clean gas of a pure air plasma and of air plasmas with an admixture of propane and toluene, respectively. The air flow was 26.0 sl/min and 12.0 sl/min, respectively, while the propane and toluene admixture was again 1.77 mg, 5.90 mg and 3.70 mg, 12.33 mg, respectively. The admixture of propane and toluene corresponds to a concentration of 900 ppm and 2800 ppm and of 405 ppm and 1400 ppm, respectively. A microwave power of 1 kW was supplied. The FTIR spectra of the clean gas of a pure air plasma show NO-, NO_2-, and N_2O_4-bands. An admixture of propane or toluene leads, additionally to the nitride oxide bands, to CO-, CO_2, and H_2O-bands but to no propane- or toluene-bands. This shows that already by a pure air plasma large amounts of NO_x are produced and that propane and toluene is degraded to carbon dioxides and water vapour.

4.3. DECOMPOSITION OF WASTE GASES

$A^2\Sigma^+ - X^2\Pi_\gamma$-transitions, which are also called the violet-, the 336 nm-, and the 306 nm-system, respectively [58]. The intensity of the CN-, NH-, and OH-bands grows with increasing concentrations of propane and toluene, which can be seen in the spectra where the propane and toluene admixture is increased to 5.90 mg and 13.33 mg, respectively, while the intensity of the NO-bands decreases slightly.

The presence of the NO-bands in the spectra of the pure air plasmas shows that NO radicals are already formed from the air molecules nitrogen and oxygen. The appearance of the CN- and NH-bands, when propane and toluene is added, shows that propane and toluene are broken up and CN and NH radicals are generated. Since OH-bands are also present in the overview spectra, OH radicals are formed, which indicates that propane and toluene are degraded to H_2O and potentially to CO and CO_2, although no CO- or CO_2-bands are observed.

Further and detailed information about the possible reaction channels are provided by the raw and clean gas analyses performed with the FTIR spectrometer which will be discussed in the following.

In Fig. 4.19 c) and d) FTIR spectra of the clean gases are shown. The FTIR spectrum of the clean gas of a pure air plasma already shows intense NO-, NO_2-, and N_2O_4-bands. This evinces that large amounts of nitride oxides are produced, which is already indicated by the presence of intensive NO-bands in the optical emission spectra. Since the plasma is close to local thermodynamic equilibrium, large amounts of nitride oxides are to be expected.

When propane or toluene is added, CO-, CO_2-, and H_2O-bands are observed in addition to the nitride oxide bands in the FTIR spectra of the clean gases. However, propane- or toluene-bands are not observed.

The intensity of the CO-, CO_2-, and H_2O-bands increases when the admixture of propane or toluene is increased. This can be seen from the FTIR-spectra when the admixture of propane or toluene is increased form 1.77 mg to 5.90 mg and from 3.70 mg to 12.33 mg, respectively. Since no propane- or toluene-bands are observed but CO-, CO_2-, and H_2O-bands are, propane and toluene are completely degraded to carbon oxides and water vapour for the regarded air flows of 12.0 sl/min and 26.0 sl/min and a supplied microwave power of 1 kW. The decomposition to water vapour is already indicated by the appearance of the OH-bands in the optical emission spectra.

Although, CN- and NH- bands are also observed in the optical emission spectra, no compounds containing CN or NH were detected in the FTIR spectra. This can be explained thereby that the CN and NH radicals, which are present in the lower part of the plasma, are degraded further up in the plasma or that they are not stable and are decompose before they are measured in the FTIR spectrometer. Furthermore, even very tiny amounts of CN or NH radicals lead to intense emission bands in optical emission spectra, since they have a high emissivity, so that the observed CN- and NH-bands can be caused by marginal amounts.

Furthermore, the destruction and removal efficiencies (DRE) of propane and toluene in dependence of the air flow and of the supplied microwave power was measured. The propane and toluene concentration in the raw and clean gases were measured with a flame ionisation detector (FID). Then the destruction and removal efficiency DRE in % was calculated by:

$$DRE = \frac{n_{raw} - n_{clean}}{n_{raw}} * 100, \qquad (4.25)$$

where n_{raw} and n_{clean} are the concentrations of propane and toluene in the raw and clean gas, respectively.

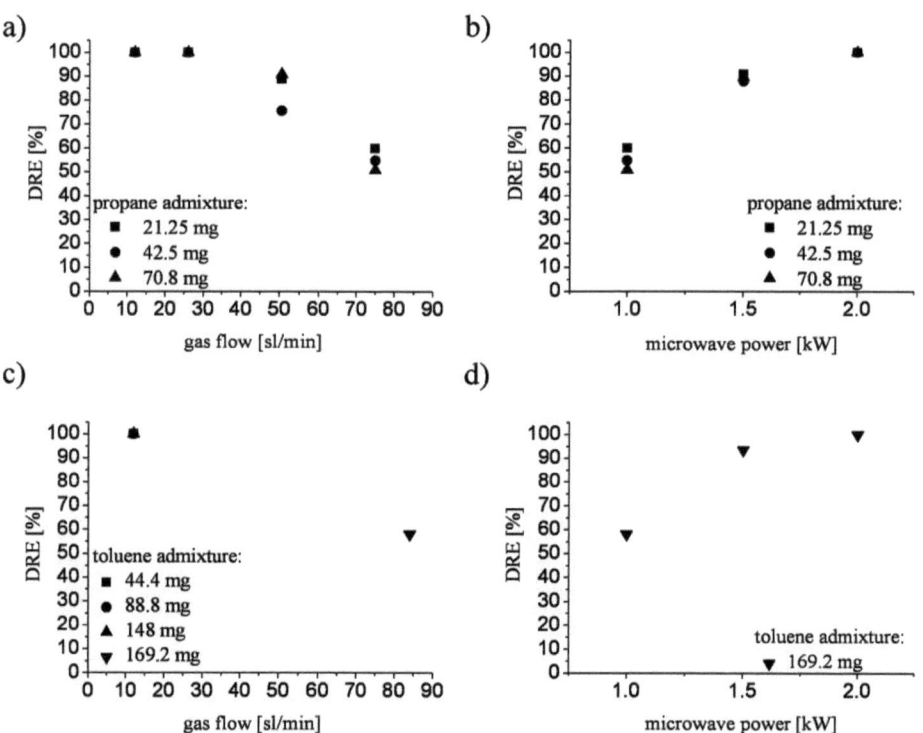

Figure 4.20: Destruction and removal efficiencies (DRE) of propane a) for a microwave power of 1 kW and different gas flows and b) for a gas flow of 75 sl/min air and different microwave powers. DRE for toluene c) for a microwave power of 1 kW and different gas flows and d) for a gas flow of 85 sl/min air and different microwave powers. The DRE decreases with an increase of the gas flow and decreases when the microwave power is increased. The dependence of the DRE on the gas flow and the supplied microwave power can explained by the extent of the plasma in dependence of the gas flow and microwave power.

The destruction and removal efficiency of propane and toluene for different gas flows and microwave powers is presented in Fig. 4.20. The diagram in Fig. 4.20 a) shows the DRE of propane for a microwave power of 1 kW in dependence of the air flow. It can be seen that for low gas

4.3. DECOMPOSITION OF WASTE GASES

flows up to 26 sl/min propane is completely degraded. An increase of the gas flow to 50 sl/min and 75 sl/min leads to a decrease of the DRE to 88.8 %..90.8 % and 50.9 %..59.9 %. The decrease of the DRE in dependence of the gas flow can be explained by the fact that the length and the diameter of the plasma decreases with increasing gas flows, as explained in section 4.2.3.

Since the length of the plasma decreases with an increasing gas flow, the dwell time of the propane in the plasma is reduced. Additionally, since the diameter also decreases in dependence of the gas flow, larger amounts of the air containing propane are able to stream past the plasma and therefore are not treated at all. Beyond these facts, the dwell time is intrinsically shortened thereby that the gas flow velocity is increased at higher gas flows since the cross section of the quartz tube stays the same.

If the microwave power is increased for a gas flow of 75 sl/min, the DRE increases to 89.5 %..90.7 % at a microwave power of 1.5 kW and at a microwave power of 2.0 kW the propane is completly degraded. This is shown in the diagram in Fig. 4.20 b). The increase of the DRE in dependence of the microwave power can again be explained by the extent of the plasma. The plasma length and diameter increase when the microwave power is increased and therefore the dwell time of the propane in the plasma is increased. Beyond that, since the diameter increases also, less propane can pass through the quartz tube without being treated by the plasma.

The same analyses of the DRE in dependence of the gas flow and of the supplied microwave power were performed for air flows containing toluene. The dependence of the DRE on the gas flow for toluene is shown in the diagram in Fig. 4.20 c). It has the same dependence like the DRE for propane: the DRE decreases when the gas flow increases.

When the microwave power is increased, the DRE for toluene in an air flow of 85 sl/min increases again and at a microwave power of 2.0 kW the toluene is almost completely degraded, as can be seen in the diagram in Fig. 4.20. The dependence of the destruction and removal efficiency on the gas flow as well as on the microwave power of toluene can again be explained by the extent of the plasma in dependence of the gas flow and the supplied microwave power. The DRE decreases with increasing gas flows, since the plasma length and diameter decreases, which leads to a shorter dwell time and larger amounts stream past the plasma. Since the length and diameter of the plasma grows with increasing microwave power the dwell time is longer and the whole gas flow streams through the plasma.

For a detailed characterisation of the formed degradation products samples were absorbed on Tenax® and analysed with a gas-phase chromatograph. These analyses were kindly performed by the Institut für Siedlungswasserbau, Wassergüte- und Abfallwirtschaft (ISWA) of der Universität Stuttgart (Institute for Sanitary Engineering, Water Quality and Solid Waste Management of the University of Stuttgart).

Gas chromatic analyses of the by-products of the degradation of propane showed that no by-products are formed during the decomposition of propane. If the degradation was not complete, propane was found as the only reaction product. Thus the FTIR and gas chromatic analyses lead

Table 4.4: Concentrations of inorganic and organic by-products in the clean gas with the degradation of toluene.

	measured concentration [mg/m^3]	threshold TA Luft [68] [mg/m^3]
organic compounds		
benzene	1..4	3..5
benzaldehyde	40..60	50
benzoic acid	50..60	55
benzoquinones	4..6	20
inorganic compounds		
CO	20..300	100
NO	1300..2300	100 as NO$_2$
NO$_2$	290..480	
N$_2$O	9..730	

to the conclusion, that propane is degraded only to carbon oxides and water vapour and no other by-products are formed with the degradation of propane by the APS.

The same gas chromatic analyses were also performed with samples of the clean gas of air flows containing toluene. In these samples many different and critical by-products, like benze, benzaldehyde, benzoic acid, and benzoquinones, were found, whose concentrations range from 1 mg..4 mg over 4 mg..6 mg to 40 mg..60 mg and 50 mg..60 mg for benzene, benzoquinones, benzaldehyde, and benzoic acid, respectively. The discovered by-products and their measured concentrations are summarised in table 4.4. Furthermore, table 4.4 summarises the measured concentrations of carbon monoxide and nitride oxides, which range from 20 mg..300 mg for CO and from 1300 mg..2300 mg, 290 mg..480 mg, and 9 mg..730 mg for NO, NO$_2$, and N$_2$O, respectively. Additionally, the TA Luft values for all these compounds are given in table 4.4. The TA Luft values are threshold values for the immission of gases according to the Erste Allgemeine Verwaltungsvorschrift zum Bundes-Immissionsschutzgesetz, Technische Anleitung zur Reinhaltung der Luft - TA Luft published by the Bundesministerium für Umwelt, Naturschutz und Reaktorsicherheit (First General Administrative Regulation Pertaining the Federal Immission Control Act - Technical Instructions on Air Quality Control - TA Luft published by the Federal Ministry for the Environment, Nature Conservation and Nuclear Safety) [68].

Like table 4.4 shows, the TA Luft threshold values for VOC are scarcely satisfied for some sets of parameters, though the threshold values for the immission of nitride oxides are distinctly exceeded. Hence the abatement of VOC with the APS, especially the abatement of toluene is questionable although DRE values of over 99 % were reached, since critical by-products and large amounts of nitride oxides are produced, which exceed the threshold values of the TA Luft.

To conclude, the studies concerning the abatement of VOC in air, here the decomposition of propane and toluene, with the APS showed that the destruction and removal efficiencies of over 99 % for both model VOC are reached. The DRE decreases with an increase of the gas flow and increases with increasing supplied microwave power. This can be explained by the extent of the plasma in dependence of the gas flow and the microwave power. The plasma diameter and length increases with increasing microwave power and decreases with an increase of the gas flow. Thus for high gas flows and low microwave powers the dwell time is short due to the small length of the plasma and not all of the gas flows through the plasma since the plasma diameter is reduced. On the other hand, if the microwave power is increased the length and diameter of the plasma is enlarged and the dwell time is increased and all of the gas is treated by the plasma.
FTIR spectroscopic analyses of the clean gas revealed that the VOC are almost completely converted to CO, CO_2 and water vapour. Furthermore these measurements showed that nitride oxides are produced in such large amounts that the TA Luft threshold values are exceeded by far. The production of nitride oxides was already indicated by the optical emission spectroscopy, since intense NO-bands in the UV range were observed. Moreover, detailed analyses of the clean gas with the gas-phase chromatograph revealed that critical by-products, like benzne, benzaldehyde, benzoic acid, and benzoquinones, are formed during the decomposition of toluene. The concentrations of these by-products scarcely satisfy the TA Luft threshold values depending on the experimental parameters. The detailed analyses of the clean gas with FTIR spectroscopic and gas chromatic studies revealed that, even though the high DRE values are reached, the application of the APS for the abatement of VOC is questionable.

4.3.2 Abatement of Perfluorinated Compounds

In the previous section 4.3.1 the suitability of the APS for the decomposition of volatile organic compounds VOC was studied. The studies showed, that propane and toluene can almost be completely degraded by the APS plasma but large amounts of nitride oxides are produced and critical by-products are formed through the abatement of toluene. Thus the application of the APS for the abatement of VOC is not that good, although the high destruction and removal efficiencies were promising. However, perfluorinated compounds (PFC), like for example CF_4 or SF_6, have a 7400 to 23000 increased green house warming potential compared to CO_2 and therefore are much more harmful than VOC. The PFC are used for etching processes in growing industry sectors, for example in semiconductor industries. Thus the decomposition of PFC is an even more urgent and important task than the decomposition of VOC nowadays.

To examine the suitability of the APS for the decomposition of PFC, the degradation of CF_4 and SF_6 in nitrogen plasmas was studied. For these abatement studies nitrogen gas flows containing

CF_4 and SF_6 were utilised, since commonly these etch gases only appear in nitrogen atmospheres. As with the studies concerning the decomposition of VOC, the fundamental principles of the decomposition of VOC through the APS were examined and therefore model waste gas mixtures consisting of nitrogen and either CF_4 or SF_6 were blended. The clean gas was analysed with a quadrupole mass spectrometer and with a FTIR spectrometer. The experimental setup is the same as for the decomposition studies of the VOC, as presented in section 4.1.2. However, during these experiments the exhaust gas system was extended by a wet vent washer. Measurements of the clean gas before the wet vent washer were performed with the quadrupole mass spectrometer and measurements behind the wet vent washer were conducted with the FTIR spectrometer. All the PFC decomposition studies were performed at the Fraunhofer Institut für chemische Technologie (ICT) (Fraunhofer Institute for Chemical Technology) in Pfinztal, since a wet vent washer was only available at the ICT. Furthermore, the plasma was again explored by optical emission spectroscopy to obtain information about the species in the plasma.

Optical emission overview spectra of a pure nitrogen plasma, of nitrogen plasmas containing CF_4 or SF_6 and of moistened nitrogen plasmas containing PFC are shown in Fig. 4.21. The spectra were recorded with the Avantes spectrometer at a radial position of $r = 0$ mm and at an axial position of about $z \approx 30..40$ mm above the resonator base. In the spectrum of a pure nitrogen plasma, N_2-bands in the UV range belonging to the $C^3\Pi_u - B^3\Pi_g$-transition, which is also called the second positive system, N_2-bands in the visible range, which belong to the $B^3\Pi_g - A^3\Sigma_u^+$-transition, the first positive system, and N_2^+-bands, belonging to the $C^2\Sigma_u^+ - X^2\Sigma_g^+$-transition, which is also called the second negative system, are observed [58].
When 16.7 mg CF_4 are added to the nitrogen gas flow, additionally to the N_2- and N_2^+-bands, CN-bands, the violet system, which belongs to the $B^2\Sigma - X^2\Sigma$ transition, are observed in the spectrum, as can be seen in Fig. 4.21a) [58]. When the nitrogen flow containing CF_4 is moistened to offer reaction partners to the CF_4 molecules, as will be explained later on, NH- (336 nm-system, $A^3\Pi - X^3\Sigma^-$-transition) and NO- (NO_β- and NO_γ-system, $B^2\Pi - X^2\Pi$- and $A^2\Sigma^+ - X^2\Pi$-transitions) bands additionally appear in the spectrum [58]. The NO-bands indicate, that nitride oxides are produced.
When 138.2 mg SF_6 is added to the nitrogen flow no additional bands appear in the spectrum, as the spectrum in Fig. 4.21 b) shows. However, if the gas flow is moistened, NO-bands are observed again, which indicates that nitride oxides are formed in the plasma.

As with the studies of the decomposition of VOC, destruction and removal efficiencies (DRE) were determined for the degradation of CF_4 and SF_6 in dependence of the microwave power and of the gas flow. The DRE was, according to equation 4.25, calculated from the concentration in the raw gas n_{raw} and from the concentration in the clean gas n_{clean}, which was determined from quadrupole mass spectroscopic analyses. The quadrupole mass spectrometer measurements were kindly performed by the Fraunhofer Institut für chemische Technologie (ICT).

4.3. DECOMPOSITION OF WASTE GASES

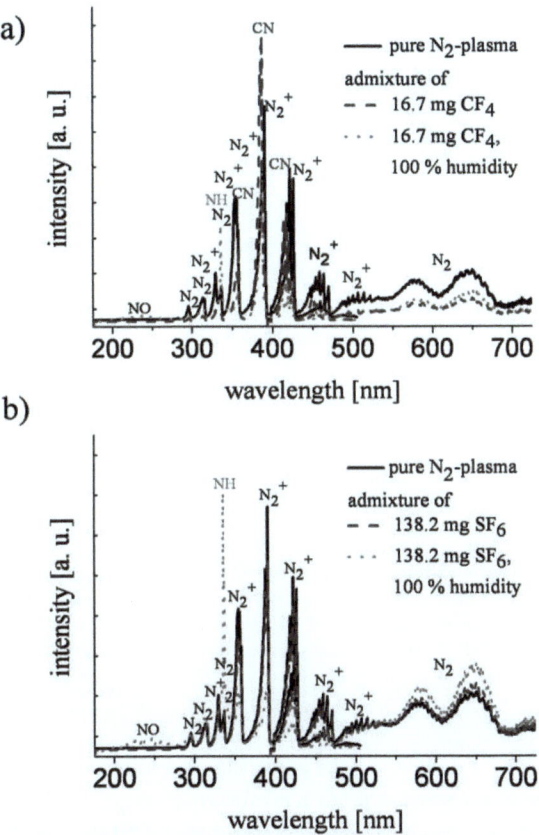

Figure 4.21: Optical emission overview spectra of nitrogen plasmas containing: a) CF_4 or b) SF_6. The spectra were recorded with the Avantes spectrometer. Three overview spectra are shown in both diagrams: a pure nitrogen plasma, a nitrogen plasma with an admixture of CF_4 or SF_6, and a humid nitrogen plasma containing CF_4 or SF_6. The nitrogen flow was 30.4 sl/min, a microwave power of 2.5 kW was supplied, and an amount of a) 16.7 mg CF_4 and b) 138.2 mg SF_6 was added.

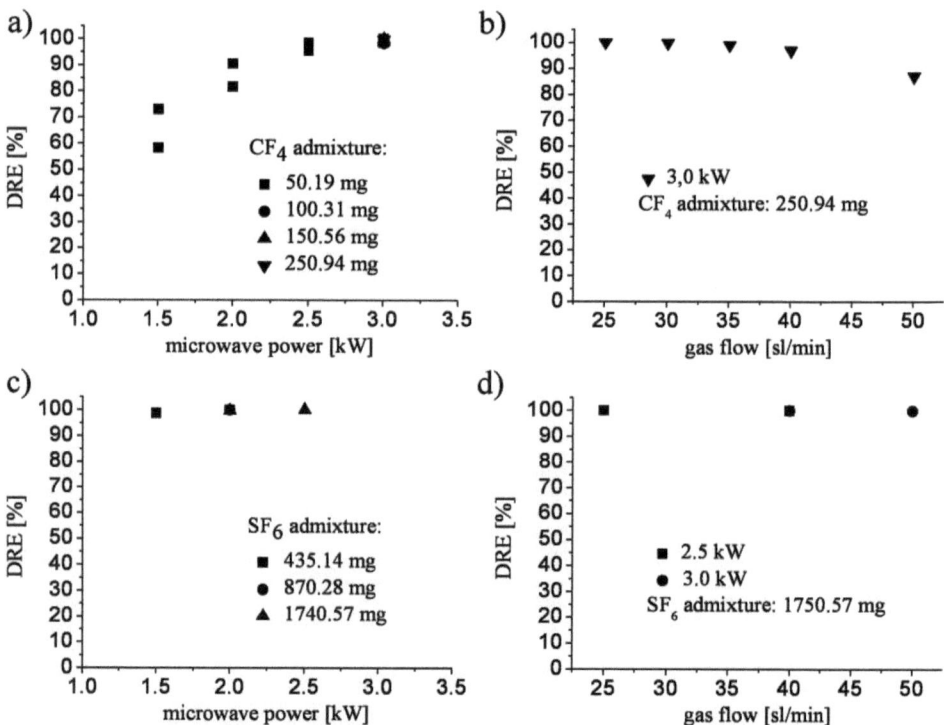

Figure 4.22: Destruction and removal efficiencies (DRE) for the degradation of CF_4 and SF_6 in dependence of the supplied microwave power and of the gas flow: a) and c) the DRE in dependence of the microwave power for a nitrogen flow of 30.4 sl/min and different admixture of CF_4 and SF_6, respectively. b) and d) DRE in dependence of the gas flow for microwave powers of 2.5 kW and 3.0 kW.

In Fig. 4.22 a) the DRE for different CF_4 admixture in a nitrogen flow of 30.4 sl/min in dependence of the microwave power are shown. At a microwave power of 1.5 kW, the DRE is only about 58.17 %..72.88 %, but if the microwave power is increased to 3.0 kW, the DRE increases, too, to 98.54 %..100 %, which corresponds to a complete degradation of CF_4. The increase of the DRE in dependence of the microwave power can again be explained by the fact that the plasma length and diameter increases when the microwave power is increased. Thus the dwell time is prolonged due to the enlargement of the plasma length. Additionally, because of the increased diameter of the plasma, all waste gas is forced to flow through the plasma and therefore all the waste gas is treated by the plasma.

Fig. 4.22 b) shows the dependence of the DRE on the gas flow for the degradation of CF_4 and a supplied microwave power of 3.0 kW. At low gas flows of 25 sl/min to 35 sl/min it can be seen that the DRE of the CF_4 reaches values of 99.97 % and 99.79 %, respectively. When the nitrogen flow is increased further to 40 sl/min or 50 sl/min the DRE decreases to 96.91 % and 86.97 %, since again

4.3. DECOMPOSITION OF WASTE GASES

the plasma extent is reduced with an increase of the gas flow.

The degradation of SF_6 was also studied. The destruction and removal efficiencies (DRE) of different SF_6 admixtures in dependence of the microwave power and of the nitrogen flow are shown in Fig 4.22 c) and d). The two diagrams show, that the DRE is independent of the supplied microwave power and gas flow in the regarded range of 1.5 kW to 3.0 kW and 25 sl/min to 50 sl/min and reaches 98.7 % to 99.97 %, which corresponds to an almost complete decomposition of SF_6. The fact that the degradation of SF_6 is almost complete can be explained thereby that the bonds within the SF_6 molecules are easier to crack compared to the bonds within the CF_4 molecules and the high temperature around the plasma is already sufficient to decompose the SF_6 through thermal degradation.

The analyses of the clean gas before the wet vent washer with the quadrupole mass spectrometer, in addition to the concentration of CF_4 and SF_6 in the clean gas, which were used to calculate the DRE, provided information about the reaction products. These analyses showed that the CF_4 molecules are converted to SiF_3, where the silicon is originated from the surrounding quartz tube. Thus to offer the C and F atoms reaction partners, the nitrogen gas flow was moistened. The measurements with the quadrupole mass spectrometer of the clean gas of humid nitrogen flows containing CF_4 showed that then the CF_4 molecules are decomposed to HF and CO_2. The wet vent washer behind the APS converted the HF molecules to calcium fluoride. The FTIR spectroscopic analyses of the clean gas behind the wet vent washer, which were kindly performed by the Institut für Siedlungswasserbau, Wassergüte- und Abfallwirtschaft (ISWA) of the Universität Stuttgart (Institute for Sanitary Engineering, Water Quality and Solid Waste Management of the University of Stuttgart), showed that only CO, CO_2, H_2O, and small amounts of NO_x could be measured. The nitride oxides are produced from the nitrogen gas flow and the humidity, which offers reaction partners to the C and F atoms.

The production of the nitride oxides was also indicated in the optical emission spectra shown in Fig. 4.21, since NO-bands in the UV range are observed when the nitrogen flow containing CF_4 or SF_6 was moistened.

To summarise, the studies concerning the degradation of PFC, here CF_4 and SF_6, in nitrogen gas flows show that destruction and removal efficiencies (DRE) of over 99 % are reached. For the decomposition of CF_4 the same dependence of the microwave power and gas flow on the DER as for the degradation of VOC was measured and can be explained by the same facts. However, the studies concerning the abatement of SF_6 showed that the DRE was independent of the analysed gas flow and microwave range of 25 sl/min..50 sl/min and 1.5 kW..3.0 kW, respectively.

Analyses of the clean gas before the wet vent washer revealed that CF_4 is converted to SiF_3 when dry nitrogen is used and to HF and CO_2 when the gas flow is moistened. The FTIR analyses behind the wet vent washer showed that only CO, CO_2, H_2O, and small amounts of nitride oxides are produced. The HF is converted to calcium fluorine in the wet vent washer. The nitride oxides are formed from the nitrogen gas flow and from the added humidity, which offers the C and F atoms reaction partners.

The PCF abatement studies showed, that high DRE are reached and that no or only little amounts of critical products, like nitride oxides, are formed during the decomposition of CF_4 and SF_6. Thus the degradation of PFC seems to be a promising application of the atmospheric pressure microwave plasma source.

In conclusion, the studies, which were performed to examine if the APS is suitable for the abatement of VOC and PFC, showed that high destruction and removal efficiencies (DRE) of over 99 % are possible. However, the DRE for propane, toluene, and CF_4 is dependent on the gas flow and the microwave power while the DRE for SF_6 is independent of the the gas flow and microwave within the examined range. The dependency of the DRE on gas flow and the microwave power the for propane, toluene, and CF_4 can be explained by the extent of the plasma, which is also influenced by these parameters. However, since on the other hand the characterisation of the plasma revealed that the tempeartures as well as the densities are almost independent of the microwave power and gas flow but the plasma volume increases with an increase of the microwave power and a decrease of the gas flow, the required plasma volume for each gas flow can be adjusted by the supplied microwave power while the same well suited conditions for a complete decompostion of the waste gases are provided.

The DRE for SF_6 is possibly independent of the gas flow and of the microwave power in the regarded range since the bonds within the SF_6 molecule are easier to crack compared to the bonds within the CF_4 molecules.

The analyses of the clean gas showed that the PFC are completely and the VOC are almost completely converted into CO, CO_2, and water vapour, and into calcium fluorine when the CF_4 is regarded. However, during the decomposition of toluene, critical by-products are formed. Additionally, large amounts of nitride oxides are produced through the degradation of VOC in air plasmas. Thus the decomposition of VOC with the APS is questionable, but the abatement of PFC seems to be a promising application of the APS.

Chapter 5
Summary and Conclusions

In the present work the development of an atmospheric pressure microwave plasma source (APS) for the abatement of waste gases as well as spectroscopic studies of the plasma and analyses concerning the decomposition of waste gases were presented.
The presented plasma torch is based on a cylindric E_{010}-mode resonator (E_{010}-APS). The power is supplied by a 2.45 GHz microwave generator (magnetron) via a rectangular waveguide. The plasma is confined in a quartz tube with an inner diameter of 26 mm. A metallic nozzle constitutes the gas inlet.

For a successful use in industrial application a straight forward handling of the plasma torch is indispensable and therefore easy ignition of the plasma as well as stable and efficient operation for different gas flows are required. To guarantee that the plasma can be ignited without any additional igniters, high electric fields must be reached in the resonator and therefore detailed information about the electric field distribution is necessary. The finite element simulations of the electric field, which were performed for this purpose, revealed that the nozzle which is used for the gas inlet and forms a coaxial structure below the cylindric resonator also acts as a resonator. Thus the plasma torch is actually based on two resonators, a cylindrical and a coaxial one, and because of that the resonant frequency of the cavity is dependent on the shape and the position of the metallic nozzle.
Moreover, the simulations showed, that a further enhancement of the electric field in the cavity can be obtained when higher mode resonators, for example E_{020}- or E_{030}-mode resonators, or special coupling elements like a slit, a $\lambda/4$-part, or a taper are used. Therefore, an E_{030}-mode based APS (E_{030}-APS) and the coupling elements were designed and manufactured.
To verify the simulation results measurements of the microwave properties of the simulated configurations were performed with a network analyser. These measurements are in excellent agreement with the simulations.

The simulations revealed and the measurements verified that, a tuning element which allows to adjust the resonant frequency of the cavity to the fixed magnetron frequency is provided by the metallic nozzle. The resonant frequency of the E_{010}-APS can be measured with a network analyser and adapted to the magnetron frequency which is also known by measurement. Then the forward powers are maximised with a three stub tuner, which is used to match the impedances and additionally provides an excellent coupling element. In this configuration a sufficiently high electric field for the ignition of a plasma in air at atmospheric pressure is already reached in the cavity when a microwave power of a few hundred watts is supplied. This plasma fills the whole quartz tube without destroying it, burns stable and the supplied microwave power is completely absorbed by the plasma.

An even better result should be achieved by an E_{030}-APS configuration since the simulations showed that these configurations provide a higher electric field. However, it turned out that due to its high quality the resonant frequency must be adjusted more precisely, which is impractical for a straight forward use of the plasma source. Furthermore, when the plasma is burning a large amount of the supplied microwave power is reflected and not absorbed by the plasma due to the sharp resonance curve.

Thus the E_{010}-APS is the optimal configuration since it provides a sufficiently high electric field for the ignition of the plasma without any additional igniters and a stable operation of the plasma which completely absorbs the supplied microwave power. Therefore, the E_{010}-APS was used for the further analyses.

The characterisation of the air plasma was performed by means of optical emission spectroscopy. The overview spectra of an air plasma are dominated by NO-bands in the UV range. Furthermore, two oxygen lines in the visible and IR range can be observed. A humidification of the plasma leads to additional OH-bands in the UV range. This $A^2\Sigma^+ - X^2\Pi_\gamma$-transition of the free OH radical was used to determine the gas rotational temperature T_{rot} which provides a good estimation of the gas temperature T_g of the plasma. The two atomic oxygen lines were used to obtain the excitation temperature T_{ex} by means of a Boltzmann plot. The excitation temperature T_{ex} gives a lower estimation of the electron density T_e. The validity of the Boltzmann plot of only tow lines was verified by Boltzmann plots of more atomic oxygen lines which were observed in an oxygen plasma. These measurements of the two temperatures spatially resolved in axial and radial direction, which were performed for three different microwave powers and air flows, revealed that the maximum gas rotational temperatures are located in the range of $T_{rot} = 3000\,\text{K}$ to $T_{rot} = 3500\,\text{K}$ and that the maximum excitation temperatures are about 2000 K higher than the gas rotational temperatures and range from $T_{ex} = 5200\,\text{K}$ to $T_{ex} = 5800\,\text{K}$. These measurements also showed that T_{rot} as well as T_{ex} are almost independent of the supplied microwave power and gas flow. However, the volume of the plasma decreases with an increase of the gas flow and increases when the microwave power is increased.

The neutral particle density and electron density, n_a and n_e, could not be directly determined from the measured optical emission spectra. Therefore, n_a was estimated by the ideal gas law and n_e by the Saha equation. The neutral particle density n_a is in the range of $2 \cdot 10^{24}\,\text{m}^{-3}..4 \cdot 10^{24}\,\text{m}^{-3}$ and the electron density reaches a maximum value of $n_e = 3 \cdot 10^{20}\,\text{m}^{-3}$ which is higher than the critical density (cutoff density) of $n_c = 7.4 \cdot 10^{16}\,\text{m}^{-3}$. Thus a maximum degree of ionisation of

$\chi \approx 10^{-4}$ is reached.

The collision frequencies which provide information about the energy transfer times, were calculated from the temperatures and densities. The collision frequencies are in the range of $\nu \approx 10^9\,\mathrm{s}^{-1}$ to $\nu \approx 10^{10}\,\mathrm{s}^{-1}$ and therefore the energy transfer times for the electrons and heavy particles are in the nano second range. Thus the electrons as well as the heavy particles obey a Maxwellian velocity distribution. Due to the mass difference the energy transfer time between the electrons and the heavy particles is only in the micro second range.

Since only the light electrons are able to follow the quickly oscillating electric field of the microwave only they are directly heated by the microwave. This results in about 2000 K higher measured temperature of the electrons. On the other hand an optimal heating of the electrons is only possible when the ratio of the electron neutral particle frequency ν_{en} and the frequency of the microwave ω is $\frac{\nu_{en}}{\omega} = 1$ which is almost reached with $\frac{\nu_{en}}{\omega} \approx 3$ for the plasma of the APS.

Since the temperatures are strongly related to $\frac{\nu_{en}}{\omega}$, the temperatures level out around an optimal value of $\frac{\nu_{en}}{\omega} = 1$. Thus a variation of the microwave power or of the gas flow cannot lead to a change of the maximum temperatures and therefore resulted in a variation of the plasma volume, as is observed.

The studies concerning the decomposition of waste gases showed that volatile organic compounds (VOC) in air plasmas as well as perfluorinated compounds (PFC) in nitrogen plasmas can be decomposed almost completely to carbon oxides, water vapour, and calcium fluorine, for CF_4, and that the destruction and removal efficiency (DRE) increases when the microwave power is increased and decreases when the gas flow is increased. This can be explained thereby that the volume of the plasma increases with an increase of the microwave power and a decrease of the gas flow which leads to longer dwell times of the waste gases in the plasma.

However, these measurements also revealed that even though DRE of over 99 % for propane and toluene were reached, the suitability of the APS for the decomposition of VOC is questionable since critical by-products and large amounts of nitride oxides are produced.

The decomposition of PFC, which have a 7400..23000 times higher green house potential compared to CO_2 and are very difficult to decompose in conventional gas or oil combustions, is a promising application of the APS since DRE of over 99 % for CF_4 and SF_6 are reached and no critical by-products and no or only very small amounts of nitride oxides are formed.

To conclude, the development of an atmospheric pressure microwave plasma source for the abatement of waste gases which provides ignition without any additional igniters as well as stable plasma operation was successful. Since DRE of over 99 % are reached for the decompostion of PFC in nitrogen plasmas and no critical by-products are formed the APS is well suited for this application. Furthermore, since the intrinsic plasma properties, like the temperatures and densities, are almost independent of the supplied microwave power and gas flow but the plasma volume is adapted, the optimal plasma volume for each required gas flow can be adjusted by the supplied microwave power while the same good conditions for a complete decomposition of the waste gases are provided.

In this work an air plasma was characterised in detail in dependence of the supplied microwave power and gas flow. However, since the APS is well suited for the abatement of PFC which commonly appear in mixtures with nitrogen, the characterisation of nitrogen plasmas would be interesting and could be part of further work. The first characterisations of nitrogen plasmas containing PFC with optical emission spectroscopy was already performed and showed that molecular as well as ion molecular bands can be observed and therefore a comprehensive characterisation of nitrogen plasmas would also provide the possibility of an independent determination of the electron density.

Furthermore, in the present work fundamental understanding of the reaction channels involved in the decomposition of waste gases was obtained and therefore for a straight forward approach only mixtures of one single harmful gas in air or nitrogen flows were studied. However, before an application in industrial processes is adopted, commonly appearing mixtures of different waste gases should be analysed.

Acknowledgements

First and foremost Prof. Dr. Uwe Schumacher has my gratitude for the excellent supervision of this work as well as for the main report and furthermore, many thanks are due to Prof. Dr. Tilman Pfau for the co-report.

Next I would like to thank Prof. Dr. Ulrich Stroth for the welcoming reception at the Institut für Plasmaforschung and Dr. Matthias Walker who introduced me to the working group Plasmatechnologie.

Furthermore, many thanks to Dr. Martin Reiser from the Institut für Siedlungswasserbau, Wassergüte- und Abfallwirtschaft (ISWA) of the Universität Stuttgart as well as to Dr. Mathias Kaiser and Dr. Lukas Alberts, who at that time were at the Fraunhofer Institut für chemische Technologie (ICT) and now are with the Muegge Electronic GmbH, for the good cooperation and the conduct of some of the analyses concerning the abatement of waste gases, which could not be preformed at the Institut für Plasmaforschung, as well as for fruitful discussions. Muegge Electronic GmbH I also would like to thank for providing some of the microwave components and for the microwave technical support.

Many more people contributed to this work through their good collaboration, fruitful discussions, their support, as well as providing helping hands in the lab, in particular Dr. Matthias Walker, Dr. Andreas Schulz, Dr. Kurt Hirsch, Dipl.-Phys. Joachim Schneider, Dipl.-Ing. Ulrich Schweitzer, Dipl.-Phys. Sebastian Enge, Dipl.-Phys. Evelyn Häberle, Dipl.-Phys. Jochen Kopecki, Dipl.-Phys. Dennis Kiesler, Simone Plog, Heinz Petto, Dipl.-Ing.(FH) Bernhard Roth, and all other colleagues.

I also wish to thank the Forschungsvereinigung für Luft- und Trocknungstechnik e.V. (FLT) and the Arbeitsgemeinschaft industrielle Forschungsvereinigungen (AiF) for partly funding this research (contract No 14248).

Finally, I would like to thank Ruben Benkmann who thoroughly and very patiently corrected my terrible English and special thanks also go to my family, my parents, my sister, and again Ruben, for the constant support and the faith in me.

List of Figures

1.1 Overview of the variety of different plasmas. 12
1.2 Dependence of the real and imaginary part of the refraction index for different collision frequencies $\frac{\nu_{en}}{\omega}$ on $\frac{n_e}{n_c}$. 20
1.3 The relative power absorption as function of the collision frequency ν_{en} normalised to ω for different $\frac{n_e}{n_c}$ ratios. 21

2.1 Comparison of different atmospheric pressure microwave plasma sources. 26
2.2 Schematic view of the atmospheric pressure microwave plasma source APS. 28
2.3 Cylindrical resonator. 29
2.4 Distribution of the norm of the electric field of E_{010}-, E_{020}-, and E_{030}-mode resonators. . 35
2.5 Distribution of the norm of the electric field of configurations with improved coupling elements. 38
2.6 Transfer properties of the coupling elements: slit, $\lambda/4$-part, and taper. 39
2.7 Eigenfrequency analyses of a cylindrical resonator with and without an inserted quartz tube. 41
2.8 Dimensions of the APS with the metallic nozzle for the simulation of the electric field distributions and Eigenfrequency analyses. 42
2.9 Electric field distribution and resonant frequency in dependence of the nozzle position. 43
2.10 Distribution of the z-component of the electric field for the coaxial and resonator mode and dependence of the resonant frequency on the nozzle position and resonator radius. 44
2.11 Schematic view of the E_{030}-APS. 48
2.12 Schematic view and photos of the three coupling elements: a slit, a $\lambda/4$-part, and a taper. 49
2.13 Schematic view of the experimental setup for the measurement of the microwave properties using a network analyser and a typical measurement. 50
2.14 Dependency of the resonant frequency on the nozzle position: Comparison simulation results - measurement. 53
2.15 Effect of the nozzle tip shape on the resonant frequency. 54

3.1 Dependency of the microwave frequency on the output power of the magnetron. 57
3.2 Commonly used experimental setup for the operation of microwave-generated plasma sources. 57
3.3 Different kinds of plasma modes in the E_{010}-APS. 59

3.4 Photos of the E_{030}-APS with a plasma. 61

4.1 Schematic view of the experimental setup, which was used to characterise the plasma of the APS and for the analyses of the decomposition of waste gases. 66
4.2 Photos of the optical setup, which was used to characterise the plasma by means of optical emission spectroscopy. 69
4.3 Measurement of the properties of the optical setup and its magnification factor. 70
4.4 Grotrian diagram for some energy levels of oxygen 72
4.5 Schematic view of the energy levels of a molecule. 76
4.6 Overview spectra of an air plasma 80
4.7 Simulated spectra of the OH-band for different temperatures and comparison between a simulated and measured spectrum. 81
4.8 Photos of the APS plasma for three different microwave powers and three different air flows. 86
4.9 Profiles of the gas rotational temperature T_{rot} spatially resolved in axial and radial direction for a microwave power of 3 kW and an air flow of 30 sl/min. 87
4.10 Profiles of the excitation temperature T_{ex} spatially resolved in axial and radial direction for a microwave power of 3 kW and an air flow of 30 sl/min. 90
4.11 Three Boltzmann plots of atomic oxygen lines, which are observed in an oxygen plasma. 91
4.12 Profiles of the electron density n_e spatially resolved in axial and radial direction for a microwave power of 3 kW and an air flow of 30 sl/min. 93
4.13 Contour plots of the gas rotational temperature T_{rot} for different microwave powers and air flows spatially resolved in axial and radial direction. 95
4.14 Contour plots of the excitation temperature T_{ex} for different microwave powers and air flows spatially resolved in axial and radial direction. 97
4.15 Contour plots of the electron density n_e for different microwave powers and air flows spatially resolved in axial and radial direction. 99
4.16 Dependence of the gas rotational temperature T_{rot} and of the excitation temperature T_{ex} on the microwave power and on gas flow inside the resonator. 101
4.17 Dependence of the gas rotational temperature T_{rot} and of the excitation temperature T_{ex} on the microwave power and on the gas flow above the resonator. 104
4.18 Dependence of the axial position where the plasma is thermalising on the microwave power and on the gas flow. 105
4.19 Optical emission spectra of a pure air plasma and of an air plasma with an admixture of propane or toluene and FTIR spectra of the clean gases. 108
4.20 Destruction and removal efficiencies of propane and toluene containing air flows for different gas flows and different microwave powers. 110
4.21 Optical emission overview spectra of a pure nitrogen plasma and of nitrogen plasmas containing CF_4 or SF_6. 115

4.22 Destruction and removal efficiencies for the degradation of CF_4 and SF_6 in dependence of the supplied microwave power and of the gas flow. 116

List of Tables

1.1 Summary of the fundamental and derived plasma parameters of atmospheric pressure microwave-generated plasmas. 23
2.1 Quality of the sole E_{010}- and E_{030}-APS and of configurations of the APS in combination with the three coupling elements. 52
4.1 Overview of the three utilised spectrometers. 68
4.2 Summary of possible methods to determine the fundamental plasma parameters by optical emission spectroscopy. 78
4.3 Summary of the determined fundamental plasma parameters of the APS plasma, applied methods, and used transitions. 84
4.4 Concentrations of organic and inorganic by-products in the clean gas with the degradation of toluene. 112

Bibliography

[1] Kyoto protocol, 1998

[2] M. Moisan, R. Pantel, J. Hubert, Propagation of a Surface Wave sustaining a Plasma Column at Atmospheric pressure, Contrib. Plasma Physics, 293 - 314, **30** (**2**), 1990

[3] Z. Zakrzewski, M. Moisan, G. Sauvé, Plasmas sustained within Microwave Circuits, in M. Moisan, J. Pelletier, Microwave excited Plasmas, Elsevier, 1992

[4] T. Fleisch, Y. Kabouzi, M. Moisan, J. Pollak, E. Castaños-Martínez, H. Nowakowska, Z. Zakrzewski, Designing an efficient microwave-plasma source, independent of operating conditions, at atmospheric pressure, Plasma Sources Sci. Technol., 173 - 182, **16**, 2007

[5] M. Moisan, G. Sauvé, Z. Zakrzewski, J. Hubert, An atmospheric pressure waveguide-fed microwave plasma torch: the TIA design, Plasma Sources Sci. Technol., 584, **3**, 1994

[6] E. A. H. Timmermans, J. Jonkers, I. A. J. Thomas, A. Rodero, M. C. Quinter, A. Sola, A. Gamero, J. A. M. van der Mullen, The behavior of molecules in microwave-induced plasmas studied by optical emission spectroscopy. 1. Plasmas at atmospheric pressure, Spectrochimica Acta Part B, 1553 - 1466, 1998

[7] E. A. H. Timmermans, I. A. J. Thomas, J. Jonkers, E. Hartgers, J. A. M. van der Mullen, D. C. Schram, The influence of molecular gases and analyses on excitation mechanisms in atmospheric microwave sustained argon plasmas, Fresenius J. Anal. Chem., 440 - 446, **362**, 1998

[8] J. Jonkers, J. M. de Regt, J. A. M. van der Mullen, H. P. C. Vos, F. P. J. de Groote, E. A. H. Timmermans, On the electron temperatures and densities in plasmas produced by the "torch à injection axiale", Spectrochimica Acta Part B, 1385 - 1392, **51**, 1996

[9] V. K. Liau, M. T. C. Fang, J. D. Yan, A. I. Al-Shamma'a, A two-temperature model for a microwave generated argon plasma at atmospheric pressure, J. Phys. D: Appl. Phys., 2774 - 2783, **36**, 2003

[10] K. M. Green, M. C. Borrás, P. P. Woskov, G. J. Flores, K. Hadid, P. Thomas, Electron Excitation Temperature Profiles in an Air Microwave Plasma Torch, IEEE Transaction on Plasma Science, 399 - 406, **29** (**2**), 2001

[11] K. Hadidi, P. P. Woskov, G. J. Flores, K. Green, P. Thomas, Effect of Oxygen Concentration on the Detection of Mercury in an atmospheric Microwave Discharge, Jpn. J. Appl. Phys., 4595 - 4600, **38**, 1999

[12] S. Y. Moon, W. Choe, H. Uhm, Y. Hwang, J. J. Choi, Characteristics of an atmospheric microwave-induced plasma generated in ambient air by an argon discharge excited in an open-ended dielectric discharge tube, Phys. Plasmas, 4045 - 4051, **9 (9)**, 2002

[13] S. Y. Moon, W. Choe, Parametric study of atmospheric pressure microwave-induced Ar/O_2 plasmas and the ambient air effect on the plasma, Phys. Plasmas, **13**, 2006

[14] S. Y. Moon, W. Choe, A comparative study of rotational temperatures using OH, O_2 and N_2^+ molecular spectra emitted from atmospheric plasmas, Spectrochimica Acta Part B, 249 - 257, **58**, 2003

[15] Y. Babou, P. Rivière, M.-Y. Perrin, A. Soufiani, Spectroscopic study of microwave plasmas of CO_2 and CO_2-N_2 mixtures at atmospheric pressure, Plasma Sources Sci. Technol., 1 - 11, **17**, 2008

[16] M. Baeva, H. Gier, A. Pott, J. Uhlenbusch, J. Höschele, J. Steinwandel, Pulsed microwave discharge at atmospheric pressure for NO_x decomposition, Plasma Sources Sci. Technol., 1 - 9, **11**, 2002

[17] M. Baeva, H. Gier, A. Pott, J. Uhlenbusch, J. Höschele, J. Steinwandel, Studies on Gas Purification by a Pulsed Microwave Discharge at 2.46 GHz in Mixtures of $N_2/NO/O_2$ at Atmospheric Pressure, Plasma Chemistry and Plasma Processing, 225 - 247, **21 (2)**, 2001

[18] A. Pott, T. Doerk, J. Uhlenbusch, J. Ehlbeck, J. Höschele, J. Steinwandel, Polarization-sensitive coherent anti-Stokes Raman scattering applied to the detection of NO in a microwave discharge for reduction of NO, J. Phys. D: Appl. Phys., 2485 - 2498, **31**, 1998

[19] A. Pott, Experimentelle und theoretische Untersuchung gepulster Mikrowellenplasmen zur Abgasreinigung in Gemischen aus Stickstoff, Sauerstoff und Stickstoffmonoxid, Inauguraldissertation, Düsseldorf, 2002

[20] Y. C. Hong, H. S. Uhm, B. J. Chun, S. K. Lee, S. K. Hwang, D. S. Kim, Microwave plasma torch abatement of NF_3 and SF_6, Phys. Plasma, **13**, 2006

[21] Y. C. Hong, H. S. Uhm, Abatement of CF_4 by atmospheric-pressure microwave torch, Phys. Plasma, 3410 - 3414, **10 (8)**, 2003

[22] H. S. Uhm, Y. C. Hong, D. H. Shin, A microwave plasma torch and its applications, Plasma Sources Sci. Technol., S26 - S34, **15**, 2006

[23] J. Mizeraczyk, M. Jasiński, Z. Zakrzewski, Hazardous gas treatment using atmospheric pressure microwave discharges, Plasma Phys. Control. Fusion, B589 - B602, **47**, 2005

[24] Y. Kabouzi, M. Moisan, J. C. Rostaing, C. Trassy, D. Guérin, D. Kéroack, Z. Zakrzewski, Abatement of perfluorinated compounds using microwave plasmas at atmospheric pressure, J. Appl. Phys., 9483 - 9496, **93** (**12**), 2003

[25] M. Moisan, Z. Zakrzewski, R. Pantel, P. Leprince, A Waveguide-Based Launcher to Sustain Long Plasma Columns Through the Propagation of an Electromagnetic Surface Wave, IEEE Transaction on Plasma Science, **3**, 203 - 214, 1984

[26] Y. Kabouzi, M. Moisan, Pulsed Microwave Discharges Sustained at Atmospheric Pressure: Study of Contraction and Filamentation Phenomena, IEEE Transaction on Plasma Science, 292 - 293, **33** (**2**), 2005

[27] Y. Kabouzi, M. D. Calzada, M. Moisan, K. C. Tran, C. Trassy, Radial contraction of microwave-sustained plasma columns at atmospheric pressure, J. Appl. Phys., **91** (**3**), 2002

[28] H. Nowakowska, M. Jasinski, J. Mizeraczyk, Electromagnetic Field Distributions in Waveguide-Based Axial-Type Microwave Plasma Source, Eur. Phys. J. D, 511 - 518, **54**, 2009

[29] S. P. Kuo, O. Tarasenko, S. Nourkbash, A. Bakhtina, K. Levon, Plasma effects on bacterial spores in a wet environment, New Journal of Physics, **8** (**41**), 2006

[30] A. Sola, M. D. Cazada, A. Gamero, On the use of the line-to continuum intensity ratio for determining the electron temperature in a high-pressure argon surface-microwave discharge, 1099 - 1110, **28**, 1995

[31] M. C. Quintero, A. Rodero, M. C. García, A. Sols, Determination of the Excitation Temperature in a Nonthermodynamic-Equilibrium High-Pressure Helium Microwave Plasma Torch, Applied Spectrocopy, 778 -784, **51** (**6**), 1997

[32] C. Yubero, M. S. Dimitrijević, M. C. García, M. D. Calzada, Using the van der Waals broadening of the spectral atomic lines to measure the gas temperature of an argon microwave plasma at atmospheric pressure, Spectrochimica Acta Part B, 169 - 176, **62**, 2007

[33] M. D. Calzada, M. Sáez, M. C. García, Characterization and study of the thermodynamic equilibrium departure of an argon plasma flame produced by a surface-wave sustained discharge, J. Appl. Phys., 34 - 39, **88** (**1**), 2000

[34] J. Margot, Studies of emission spectra in helium plasmas at atmospheric pressure and local thermodynamical equilibrium, Physics. Plasma, 2525 - 2531, **8** (**5**), 2001

[35] Y. Kabouzi, M. D. Calzada, M. Moisan, C. Trassy, Gas temperature in contracted atmospheric pressure discharges sustained in cylindrical tubes by microwaves at 2450 MHz, Proceedings of 6'th European Conference on Thermal Plasma Processes, Strasbourg, 2000

[36] E. Tratarova, F. M. Dias, E. Felizardo, C. M. Ferreira, B. Gordiets, Microwave plasma torches driven by surface waves, Plasma Sources Sci. Technol., 1 - 7, **17**, 2008

[37] E. Iordanova, J. M. Palomares, A. Gamero, A. Sola, J. J. A. M. van der Mullen, A novel method to determine the electron temperature and density from the absolute intensity of line and continuum emission: application to atmospheric microwave induced Ar plasmas, J. Phys. D: Appl. Phys., 1 - 12, **42**, 2009

[38] M. Christova, E. Castaños-Martinez, M. D. Calzada, Y. Kabouzi, J. M. Lique, M. Moisan, Electron Density and Gas Temperature from Line Broadening in an Argon Surface-Wave-Sustained Discharge at Atmospheric Pressure, Applied Spectroscopy, 1032 - 1037, **58** (**9**), 2004

[39] C. Tendero, C. Tixier, P. Tristant, J. Desmaison, P. Leprince, Atmospheric pressure plasmas: A review, Spectrochimica Acta Part B, **61**, 2 -30, 2006

[40] Q. Jin, C. Zhu, M. W. Borer, G. M. Hieftje, A microwave plasma torch assembly for atomic emission spectrometry, Spectrochimica Acta Part B, **46**, 417 - 430, 1991

[41] D. Kiesler, Untersuchung zur Skalierbarkeit eines Mikrowellen-Plasmabrenners bei Atmospährendruck, Diplomarbeit, Institut für Plasmaforschung, Stuttgart, 2008

[42] U. Schumacher, Fusionsforschung - Eine Einführung, wissenschaftliche Buchgesellschaft Darmstadt, 1993

[43] U. Stroth, Einführung in die Plasmaphysik, Skript zur Vorlesung, 2007

[44] G. Janzen, Plasmatechnik, Hüthig, 1992

[45] A. P. Thorne, Spectrophysics, Chapman and Hall, 1988

[46] Bergmann-Schäfer, Lehrbuch der Experimentalphysik, Band 5: Vielteilchensysteme, 1992

[47] K. Behringer, Einführung in die Plasmaspektroskopie, Skript zur Vorlesung, 2001

[48] K. Behringer, Spektroskopische Diagnostik von Nichtgleichgewichtsplasmen, Skript zur Vorlesung, 2001

[49] M. Capitelli, C. M. Ferreira, B. F. Gordiets, A. I. Osipov, Plasma Kinetics in Atmospheric Gases, Springer, 2000

[50] J. Jackson, Classical Electrodynamics, John Wiley and Sons, 1962

[51] COMSOL Multiphysics^{TM} user manual

[52] M. Leins, K.-M. Baumgärtner, M. Walker, A. Schulz, U. Schumacher, U. Stroth, Studies on a Microwave-Heated Atmospheric Plasma Torch, Plasma Process. Polym., **4**, S493 - S497, 2007

[53] M. Leins, K.-M. Baumgärtner, M. Walker, A. Schulz, U. Schumacher, U. Stroth, Studies on a microwave-heated plasma torch, in Proc. 28 ICPIG, Prague, 2007

[54] M. Leins, A. Schulz, M. Walker, U. Schumacher, U. Stroth, Development and Characterization of an Atmospheric-Pressure Microwave Plasma Torch, IEEE Trans. Plasma Sci., **36**(4), 982, 2008

[55] E. Pehl, Mikrowellentechnik, Band 1: Wellenleitungen und Leitungsbausteine, Hüthing, 1984

[56] H. Haken, H. C. Wolf, Atom- und Quantenphysik, Einführung in die experimentellen und theoretischen Grundlagen, Springer, 2004

[57] H. Haken, H. C. Wolf, Molekülphysik und Quantenchemie, Einführung in die experimentellen und theoretischen Grundlagen, Springer, 2003

[58] R. W. B. Pearse, A. G. Gaydon, The Identification of Molecular Spectra, Chapman and Hall, 1976

[59] D. Kiselmann, Non-LTE effects on oxygen abundance determinations for solar-type stars, Astronomy and Astrophysics, **245**, L9 - L12, 1991

[60] S. Pellerin, J. M. Cromier, F. Richard, K. Musiol, J. Chapelle, A spectroscopic diagnostic method using UV OH band spectrum, J. Phys. D: Appl. Phys., 726 - 739, **29**, 1996

[61] C. de Izarra, UV OH spectrum used as a molecular pyrometer, J. Phys. D: Appl. Phys., 1697 - 1704, **33**, 2000

[62] J. Happold, Ortsaufgelöste spektroskopische Untersuchung an einem mikrowellenerzeugten Plasma, Diplomarbeit, Institut für Plasmaforschung, Stuttgart, 2005

[63] J. Happold, P. Lindner, B. Roth, Spatially resolved temperature measurements in an atmospheric plasma torch using the $A^2\Sigma^+$, $\nu' = 0 \rightarrow X^2\Pi, \nu = 0$ OH band, J. Phys. D: Appl. Phys., 3615 - 3620, **39**, 2006

[64] G. H. Dieke, H. M. Crosswhite, The Ultraviolet Bands of OH, J. Quant. Spectrosc. Radiat. Transf., 97 - 199, **2**, 1962

[65] E. Felizardo, E. Tatarova, F. M. Dias, C. M. Ferreira, B. Gordiets, Microwave Air-Water Plasma Torch - Experiment and Theory, in Proc. 36th EPS Confernece on Plasma Phys. Sofia, **33E**, P-2.111, 2009

[66] H. Peng, C. Xi, L. He-ping, On the correct form of the Saha equation for tow-temperature plasmas, Chin. Phys. Lett., 193 - 195, **3**, 1999

[67] L. Alberts, M. Kaiser, M. Leins, M. Reiser, Über die Möglichkeit des Abbaus von C-haltigen Abgasen mit atmosphärischen Mikrowellen-Plasmen, in. Proc. UMTK 2008 Neue Entwicklungen bei der Messung und Beurteilung der Luftqualität, Nürnberg, 2008

[68] Erste Allgemeine Verwaltungsvorschrift zum Bundes-Immissionsschutzgesetz, Technische Anleitung zur Reinhaltung der Luft - TA Luft, Bundesministerium für Umwelt, Naturschutz und Reaktorsicherheit, 24. Juli 2002

[69] L. Alberts, M. Kaiser, M. Leins, M. Reiser, Abbau von fluorhaltigen Kohlenwasserstoffen durch atmosphärische Mikrowellenplasmen, in Proc. UMTK 2008 Neue Entwicklungen bei der Messung und Beurteilung der Luftqualität, Nürnberg, 2008

[70] M. Leins, A. Schulz, M. Walker, U. Schumacher, U. Stroth, M. Reiser, L. Alberts, M. Kaiser, K.-M. Baumgärtner, Entwicklung und Charakterisierung einer Mikrowellen-Plasmaquelle bei Atmosphärendruck für den Abbau von VOC-haltigen Abgasen, in Proc. UMTK 2008 Neue Entwicklungen bei der Messung und Beurteilung der Luftqualität, Nürnberg, 2008

[71] M. Leins, L. Alberts, M. Kaiser, M. Walker, A. Schulz, U. Schumacher, U. Stroth, Development and Characterisation of a Microwave-heated Atmospheric Plasma Torch, Plasma Process. Polym., S227 - S232, **6**, 2009

Die VDM Verlagsservicegesellschaft sucht für wissenschaftliche Verlage abgeschlossene und herausragende

Dissertationen, Habilitationen, Diplomarbeiten, Master Theses, Magisterarbeiten usw.

für die kostenlose Publikation als Fachbuch.

Sie verfügen über eine Arbeit, die hohen inhaltlichen und formalen Ansprüchen genügt, und haben Interesse an einer honorarvergüteten Publikation?

Dann senden Sie bitte erste Informationen über sich und Ihre Arbeit per Email an *info@vdm-vsg.de*.

Sie erhalten kurzfristig unser Feedback!

VDM Verlagsservicegesellschaft mbH
Dudweiler Landstr. 99　　　　　Telefon +49 681 3720 174
D - 66123 Saarbrücken　　　　　Fax　　　+49 681 3720 1749

www.vdm-vsg.de

Die VDM Verlagsservicegesellschaft mbH vertritt

Printed by Books on Demand GmbH, Norderstedt / Germany